High Definition Television
A Bibliography

High Definition Television
A Bibliography

William Saffady

Meckler

Westport • London

Library of Congress Cataloging-in-Publication Data

Saffady, William, 1944-
  High definition television : a bibliography / William Saffady.
    p. cm.
  ISBN 0-88736-422-5 (alk. paper) : $
  1. High definition television--Bibliography. I. Title.
Z7711.S24   1990
[TK6679]
016.621388--dc20   90-30802
                                                    CIP

British Library Cataloguing in Publication Data

Saffady, William, *1944-*
  High definition television : a bibliography.
  1. Television, high definition. Bibliographies
  I. Title
  016.621388

  ISBN 0-88736-422-5

Meckler Corporation, 11 Ferry Lane West, Westport, CT 06880.
Meckler Ltd., Grosvenor Gardens House, Grosvenor Gardens,
  London SW1W 0BS, U.K.

Printed on acid free paper.
Printed and bound in the United States of America.

Contents

Introduction

This bibliography contains over 1,550 citations to publications dealing with High Definition Television (HDTV) technology, products, applications, and markets. It is designed as a "reading list" for information management and communication professionals, as well as a starting point for researchers interested in HDTV technology. The listing is divided into three parts:

-- PART ONE contains citations to books, reports, journal articles, and magazine articles. Many of the cited publications are highly technical in character and will prove of greatest interest to scientists, engineers, and others responsible for the development and/or evaluation of HDTV systems. Representing the core of published literature about HDTV technology, these citations reflect the work of researchers, broadcasters, information specialists, and others in many countries. For the convenience of American readers, titles of articles published in non-English languages are given in English translations whenever possible.

-- PART TWO contains citations to papers and presentations in published conference proceedings. The titles of the papers are presented in English; for papers in non-English languages that utilize the Roman alphabet, the titles of publication in which they appear are given in both the original language and in an English translation. It should be noted that some papers delivered at professional conferences are subsequently published, in identical or edited version, in journals. Such publications are cited in PART ONE of this bibliography.

-- PART THREE contains citations to selected newspaper and news magazine articles dealing with HDTV. Since 1987, new television technologies have been the subject of considerable public discussion. The citations provided in PART THREE of this bibliography reflect the scope and content of that discussion. Most of the articles are non-technical in nature and quite brief; some deal with such controversial issues as government subsidies for HDTV and the role of HDTV as a general stimulant to the electronics industry in the United States and elsewhere. Although they have been or may be supplanted by newer publications, these citations to news articles will provide background information for readers who are examining the broader implications of High Definition Television and related technologies. It must be emphasized that this listing is selective; the citations presented here represent perhaps 10 percent of the news stories published about HDTV in the late 1980s. To provide the broadest perspective, citations to U.S. national, U.S. regional, and international publications are included.

As the foregoing discussion suggests, the publications listed in this bibliography range from brief commentary about the potential of high definition television to highly specialized and detailed treatments of HDTV system components. The scope of HDTV systems is interpreted broadly. The cited publications deal with a varied group of topics, including technological fundamentals, broadcasting methodologies, HDTV display and recording equipment, HDTV applications, and the national and international economic and policy implications of new television technologies. Citations dealing with related technologies, such as Improved Definition Television and Enhanced Definition, are also included.   This bibliography reflects works published through 1989. Reflecting the intense recent interest in HDTV, most of the references date from 1988 and 1989, although a number of older technical articles are included.

William Saffady
School of Information Science and Policy
State University of New York at Albany

Part One: Books, Reports, Journals, and Magazines

Abe, H. et al. Magnetic recording of a high-definition television signal. SMPTE Journal; 1981 March; 90(3): 192-5.

Abe, M. et al. A high-speed digital filter LSI for video signal processing. IEEE Journal of Solid-State Circuits; 1987 June; SC-22(3): 396-402.

Adachi, O. et al. 43-inch super-large color CRT. National Technical Reports (Japan); 1987 April; 33(2): 168-75.

Akiyama, M.; Minami, K.; Seta, M. Telecommunication-broadcasting integrated network and minimum cost construction of its subscriber network. Transactions of the Institute of Electronic & Communication Engineers of Japan Part B (Japan); 1985 December; J68B(12): 1349-58.

Allan, R. ISSCC: special purpose chips. Electronic Design; 1986 February 20; 34(4): 119-23.

Alster, N. TV's high-stakes, high-tech battle: it's over HDTV, the biggest thing since color; both Japan and Europe are well ahead; can the U.S. electronics industry catch up? Should the government help? If so, how? Fortune; 1988 October 24; 118(9): 161-70.

Amor, H. High defintion television. Fernseh und Kino-Technik (Germany); 1982 March; 36(3): 99-102.

Amorese, P.; Bloomfield, J. A slew of standards for camera systems. ESD: The Electronic System Design Magazine; 1988 March; 18(3): 94-98.

Ando, K. et al. A 54-inch (5:3) high-contrast high-brightness rear-projection display for high-definition TV. Proceedings of S.I.D.; 1985; 26(4): 315-21.

Ando, T. Trends of low light level imaging technology. Journal of the Institute of Television Engineers of Japan (Japan); 1988 August; 42(8): 775-9.

Andrews, W. Distinctions blur between DSP solutions. Computer Design; 1989 May 1; 28(9): 86-95.

Anonymous. Challenge of broadcasting technology is clear. JEE, Journal of Electronic Engineering; 1989 January; 26(265): 86-98.

Antipin, M.V. A satellite high definition TV system. Tekhnika Kino i Televideniya (USSR); 1985 July; (7): 65-6.

Anus, H. European manufacturers' strategy for new TV services in Europe. Bulletin IDATE (France); 1986 November; (25): 629- 42.

Aoyama, M.; Hatanaka, Y. Electron gun for high definition TV camera tubes. Bulletin of the Research Institute of Electronics, Shizuoka University (Japan); 1984; 19(2): 35- 45.

Araki, Y.; Kurosaki, T.; Urano, J. Experiments on enhanced TV system compatible with NTSC. IEEE Transactions on Consumer Electronics; 1988 August; 34(3): 452-9.

Arlen, G.H.; Prince, S.; Trost, M. Tomorrow's TVs; a review of new TV set technology, related video equipment, and potential market impacts, 1987/1995. Washington, D.C.: National Association of Broadcasters; 1987.

Armbruster, H. Future developments in the telecommunications field: universal broadband ISDN networks. Nachrichtentechnische Zeitschrift; 1987 August; 40(8): 564- 9.

Arragon, J.P. et al. HDTV picture transmission compatible with D2-MAC/packet format. Acta Electronica (France); 1985; (1- 2): 19-32.

Asatami, K.; Sato, K.; Maki, K. Fibre optic analogue transmission experiment for high-definition television signals using semiconductor laser diodes. Electronic Letters (Great Britain); 1980 July 3; 16(14): 536-8.

Ashibe, M.; Mitsuhashi, K.; Tsuruta, S. A bandwidth compression technique of HDTV signals. NEC Technical Journal (Japan); 1988 September; 41(9): 54-9.

Ashizaki, S. et al. Direct-view and projection CRTs for HDTV.

IEEE Transactions on Consumer Electronics; 1988 February; 34(1): 91-9.

Baack, C.; Kaiser, W. eds. Wege zu besseren fernsehbildern; vortrage des vom 21.-22. Januar 1987 in Muncheabgehaltenen Kongresses (Ways towards high definition TV: proceedings of the Congress held in Munich, January 21-22, 1987). Berlin, New York: Springer-Verlag; 1987.

Babcock, W.E.; Wedam, W.F. Practical considerations in the design of horizontal deflection systems for high-definition television displays. IEEE Transactions on Consumer Electronics; 1983; CE-29(3): 334-49.

Bachus, E.-J. et al. Components of a ten-channel coherent HDTV/TV distribution system. Proceedings of SPIE - The International Society for Optical Engineers; 1988; 864: 82-7.

Baker, S. Euroview. HDTV. Nothing but trouble? Cable Satellite Europe (Great Britain); 1988 June; (6): 28-9, 32-3, 35.

Baker, S. MAC comes into clearer focus (European HDTV). Cable Satellite Europe (Great Britain); 1988 November; (11): 46-8.

Baldwin, J.L.E. Enhanced television-a progressive experience. SMPTE Journal; 1985 September; 94(9): 904-13.

Bannai, T. et al. Wide band video signal recorder having level and linearity corrector. IEEE Transactions on Consumer Electronics; 1986 August; CE-32(3): 268-73.

Barbieri, G. Digital encoding of HDTV signals for applications in satellite transmission. Elletronica e Telecomunicazioni (Italy); 1988; 37(1): 29-33.

Barlas, S. Era of HDTV sending mixed signals to marketers. Marketing News; 1989 May 22; 23(11): 8-9, 23.

Barrett, A.C. The potential of fiber optics to the home: a regulator's perspective. Public Utilities Fortnightly; 1989 January 19; 123(2): 14-17.

Basile, C. Channel matching techniques for 2-channel television. IEEE Transactions on Consumer Electronics; 1987 August; CE- 33(3): 154-61.

Bechis, D.J. Guns and yokes - computer aid for modern designs. Information Display; 1988 June; 4(6): 10-14.

Belan, J.P. The NHK's high-definition television system and commercial equipment. Revue Radiodiffusion et Television (France); 1988 July/September; 22(102): 12-18.

Belforte, P. et al. Program selector for digital high quality television. CSELT Technical Reports (Italy); 1987 June; 15(1): 243 51.

Belforte, P. et al. Program selector for digital high quality television. CSELT Technical Reports (Italy); 1987 June; 15(4): 243-51.

Bencini, L. High definition television: outline of the international situation and domestic (Italian) aspects (Televisione ad alta difinizione: panoramica della situazione internationale i riflessi nazionali). Note Recensioni Notizie (Italy); 1986 April-June; 35(2): 71-74.

Bergmann, H. The new HD-MAC television transmission system. Radio Fernsehen Elektronik (East Germany); 1989; 38: 185-7.

Bergmann, H. Pick-up tube for higher resolution. Radio Fernsehen Elektronik (East Germany); 1986; 35(4): 227-8.

Bergmann, H. Picture receiver valves (television). Radio Fernsehen Elektronik (East Germany); 1988; 37(1): 15-19.

Berkman, B.N. The best and brightest speak out. Electronic Business; 1989 June 12; 15(12): 28-37.

Berlin, K. HDTV proposals: where do we stand? CED; 1989 April; 15(4): 22, 24, 27-8, 30-1.

Bernard, J. High definition TV. Radio-Electronics; 1987 August; 58(8): 48-51.

Besier, H. et al. Experimental system for the transmission of signals for narrow-and broadband serivces via glass fiber subscriber lines (Ein versuchssystem fuer die uebertragung von signalen fuer schmalband-und breitbanddienste ueber glasfaseranschlussleitungen). Fermelde-Ingenieur (Germany); 1986 April; 40(4): 1-36.

Besier, H. et al. An experimental system for signal transmission of narrowband and broadband services over optical fibre link. Der Fernmelde-Injenieur (Germany); 1986 April; 40(4): 1-36.

Bidmead, C. CD-ROM: the technology for a storage revolution. Which Computer?; 1989 July: 70-3.

Biemond, J. ed. Fifth workshop on multidimensional signal processing. Signal Processing (Netherlands); 1988 October; 15(3): 227-350.

Billotet-Hoffmann, C.; Sauerburger, H. Problems of field rate

conversions in HDTV. Proceedings of SPIE International
Society for Optical Engineering; 1986; 594: 49-56.

Bock, G.; Evers, R.; Riemann, U. Ways to better television
pictures (HDTV). Fernseh und Kino-Technik (West
Germany); 1987 December; 41(12): 563-4.

Boehm, R.J. Bringing fiber to the subscriber. Telephone
Engineering and Management; 1987 December 1; 91(23
(Part 1)): 92-7.

Bohn, J.; Heister, H.; Puhler, H.G. High definition television
systems. Requirements to be met by individual parts.
Funkschau (West Germany); 1982 November 12; (23): 72-4.

Bolle, G. Notes on new television standards. Bosch Technische
Berichte (West Germany); 1985 7; 6(237-41).

Borner, R. 3D TV projection. Electronics & Power; 1987 June;
33(6): 379-82.

Borner, R. Progress in projection of parallax-panoramagrams onto
wide-angle lenticular screens. Proceedings of SPIE The
International Society of Optical Engineers; 1987; 761:
35- 43.

Borner, R. Spatial images by the lenticular scanning process:
three-dimensional without spectacles. Funkschau (West
Germany); 1987 January 16; (2): 36-9.

Bourguignat, E. The psychovisual bases of picture improvement.
Revue Radiodiffusion et Television (France); 1985
June/August; 19(88): 6-15.

Bourguinat, E. Psychovisual bases for television pictures
enhancement. Proceedings of SPIE. The International Society
of Optical Engineers; 1986; 594: 57-64.

Bourguignat, E. The visual quality of high definition television.
Revue Radiodiffusion et Television (France); 1988
July/September; 22(102): 3-11.

Boyer, R. Compatible evolution towards HDTV. Onde Electronique
(France); 1989 July/August; 69(4): 26-34.

Boyer, R. The EUREKA high-definition television project. Bulletin
IDATE (France); 1986 November; (25): 644-7.

Braun, R.-P.; Ludwig, R.; Molt, R. Coherent optical 10 channel
wideband transmissions with optical transient wave
amplifier. Nachrichtentechnische Zeitschrift (West Germany);

1986 December; 39(12): 804-8.

Breide, S. Transmission of MAC-compatible colour-difference signals in a future HDTV system. Rundfuntechnische Mitteilungen (West Germany); 1988 July/August; 32 4 173-9.

Bretl, W.E. 3 multiplied by NTSC - a 'leapfrog' production standard for HDTV. SMPTE Journal; 1989 March; 98(3): 173-78.

Bretl, W.E. 3XNTSC - a 'leapfrog' production standard for HDTV. IEEE Transactions on Consumer Electronics; 1988 August; 34(3): 484-492.

Brett, M.D. A multi-standard MAC decoder. Electronic Technology (Great Britain); 1989 February; 23(2): 36-8.

Brilliantov, D.P.; Khleborodov, V.A. On choosing a HDTV standard: state-of-the-art HR color picture tubes. Tekhnika Kino i Televideniya (USSR); 1987 June; (6): 39-46.

Brody, H. TV: The push for a sharper picture. High Technology Business; 1988 April; 8(4): 25-29.

Bruch, W. From 625 lines to HDTV. Funkschau (West Germany); 1988 October 21; (22): 20-1 supplement.

Buchwald, W.P. Resolution increase in multichip colour cameras (TV). Fernseh und Kino-Technik (West Germany); 1985 April; 39(4): 169-74.

Bucken, R. Eureka 95: first HDTV camera with progressive sampling. Fernseh und Kino-Technik (West Germany); 1988 August; 42(8): 368-70.

Bucken, R. HDTV: experiments in Berlin. Funkschau (West Germany); 1989 June 2; (12): 50-4.

Bucken, R. High-definition television (HDTV): quarrels about the future of television. Funkschau (West Germany); 1986 August 1; (16): 87-91.

Bucken, R. The super-television-high-definition TV (HDTV): ready in Japan. Sound (Switzerland); 1987 November; 10(11): 42-4.
Bucken, R. Uncertain television future: with D2-MAC to HDTV? Funk-Technik (Germany); 1986 December; 41(12): 509-14.

Buj, J. Technological incompatibility in the electronics industry. Revista Espana Electronico (Spain); 1987 March; (388): 20-8.

Burns, R..W. Seeing by electricity (history of television). IEE

Proceedings A (Great Britain); 1986 January; 133(1): 27-37.

Bursky, D. Bright lights of New York overshadowed by ISSCC. Electronic Design; 1989 January 12; 37(1): 35-8.

Bykhovsky, V.A. Principles of building a terrestrial TV network in the 12 GHz band. Radio-Television (Czechoslovakia); 1987; 37(4): 29-33.

Bytheway, D.L. Distribution switcher for HDTV. SMPTE Journal; 1989 June; 98(6): 425-33.

Cadot, E.; Baucher, P.-Y. Cinema and HDTV: their future relations? Revue Radiodiffusion et Television (France); 1988 October; 22(103): 21-5.

Caillot, J. HDTV-strategy for a world standard. Funkschau (West Germany); 1988 August 26; (18): 33-5.

Carey, J. Washington Inc.? Business Week; 1989 June 16; (3110): 40-41.

Carlson, C.R.; Bergen, J.R. Perceptual considerations for high-definition television systems. SMPTE Journal; 1984 December; 93(12): 1121-6.

Catier, E. Display systems: cathode-ray tubes. Toute Electronique (France); 1985 May; (504): 34-42.

Cavallerano, A.P. Decomposition and recombination of a wide aspect ratio image for ENTSC two-channel television. IEEE Transactions on Consumer Electronics; 1987 August; CE-33 (3): 162-72.

Cawthorne, N. World developments in stereo sound TV. Institute of Broadcasting Systems & Operations (Great Britain); 1985 July/August; 8(7): 20-1.

Chang, K.Y.; Hara, E.H. Fiber-optic broad-band integrated distribution: ELIE and beyond. IEEE Journal on Selected Areas of Communication; 1983 April; SAC-1(3): 439-44.

Chiariglione, L. ed. Signal processing of HDTV; proceedings of the second International workshop on signal processing of HDTV. Amsterdam, New York: North Holland; 1988.

Childs, I. HDTV-putting you in the picture. IEE Review (Great Britain); 1988 July 14; 34(7): 261-5.

Childs, I.; Melwig, R. HDTV standard-setting on psycho-physical bases. EBU Review of Technology (Belgium); 1986 October; (219): 281-7.

Childs, I.; Roberts, A. The compatibility of HDTV sources and replay characteristics with present TV systems. Fernseh und Kino-Technik (West Germany); 1985 May; 39(5): 218-23.

Childs, I.; Roberts, A. The compatibility of HDTV sources and displays with present day television systems. Journal of the Institute of Electronic & Radio Engineers (Great Britain); 1985 October; 55(10): 348-52.

Chin, D. et al. The Princeton Engine: a real-time video system simulator. IEEE Transactions on Consumer Electronics; 1988 May; 34(2): 285-97.

Chitnis, E.V.; Karnik, K.S.; Jain, G.C. Electronics for teaching and mass communication. Electronic Information and Planning (India); 1983 June; 10(9): 593-600.

Choquet, B. The MUSE system. Revue Radioduffusion et Television (France); 1988 July/September; 22(102): 28-34.

Clark, R. et al. The laser marketplace 1989. Lasers & Optronics; 1989 January; 8(1): 37-49.

Coates, C. It just seems some axioms never will die-network TV. Advertising Age; 1981 November 9; 52(47): S-16, S-20.

Cohen, A. Look sharp, suppliers: HDTV will be a great show. Electronic Business; 1988 April 1; 14(7): 28, 30.

Colaitis, M.-J. The HDMAC system: principles and base-band aspects. Onde Electronique (France); 1989 July/August; 69(4): 13-19.

Colaitis, M.-J. The HDMAC system: principles and base-band aspects. Revue Radiodiffusion et Television (France); 1988 July/September; 22(102): 19-23.

Crawford, D. 3DTV -- a view of the future. British Telecom Journal (Great Britain); 1987; 8(1): 45-9.

Crawford, D.I. High definition television-parameters and transmission. British Telecom Technology Journal (Great Britain); 1987 October; 5(4): 76-83.

Cripps, D. Understanding the MAC objective (HDTV). International Broadcasting (Great Britain); 1988 December; 11(10): 36, 40.

Crutchfield, B. HDTV delivery to the consumer. IEEE Transactions on Broadcasting; 1987 December; BC-33(4): 184-7.

Crutchfield, E.B. Advanced television terrestrial broadcast project. IEEE Transactions on Consumer Electronics; 1987 August; CE-33(3): 142-5.

Cutts, D. No CA ... no comment (conditional access TV). Cable

Satellite Europe (Great Britain); 1988 April; (4): 19-21, 23.

D'Amato, P. The origination of HDTV programmes. EBU Review of Technology (Belgium); 1986 October; (219): 288-96.

D'Amato, P. The production of high definition television programmes. Elletronica e Telecomunicazioni (Italy); 1986 September/October; 35(5): 221-9.

Daubney, C. High definition television - the next step? An independent broadcasting perspective. Image Technology (Great Britain); 1986 September; 68(9): 441, 443, 460.

Day, A.; Kirton, P.; Park, J. Broadband ISDN and fast packet switching. Telecommunication Journal of Australia (Australia); 1989; 39(1): 29-36.

Day, S. Advanced television systems-what they offer. Broadcast Technology (Canada); 1988 June; 13(9): 50-2.

De Selys, G.; Maesschalk, A. Europe's audiovisual challenge. Europe; 1989 April; (285): 28-30.

De Vries, J.P. Technical developments in television broadcasting systems. PTT Bedrijf (Netherlands); 1985 March; 23(2-3): 112-16.

Dean, R. Broadcast cameras of the future. International Broadcasting Systems & Operations (Great Britain); 1985 December; 8(12): 5,8,34.

Dean, R. HDTVS looms large on camera horizon. International Broadcasting (Great Britain); 1988 December; 11(10): 19-20.

Dean, R. Telecine-the state of the manufacturers. International Broadcasting (Great Britain); 1988 September; 11(7): 90, 92- 3.

DeMarsh, L.E.; Firth, R.R.; Sehlin, R.C. Scanning requirements for motion-picture post-production. SMPTE Journal; 1985 September; 94(9): 921-4.

Deutrich, J. Further developments in high quality television pictures in Europe. Nachrichtentechnische Zeitschrift (West Germany); 1988 July; 41(7): 400-4.

DeWitt, R. The big question is, what kind of impact will ISDN have on your corporate network? Communication News; 1987 January; 24(1): 44-8.

DiGiulio, E.M. SMPTE study group on 30-frame film rate: final committee report on the feasibility of motion-picture frame-rate modification to 30 frames/sec. SMPTE Journal; 1988

May; 97(5): 404-8.

Dobbie, W.H. A DBTV system for optimum bandwidth efficiency. IEEE Transactions on Consumer Electronics; 1987 February; CE- 33(1): 58-64.

Donahue, H.C. Choosing the TV of the future. Technology Review; 1989 April; 92(3): 30-40.

Drury, G.M. Television in the eighties. BKSTS Journal (Great Britain); 1982 August; 64(8): 378-84.

Ebner, A.; Schafer, R. Aspects of conversion from HDTV standards to current television standards. Rundfunktechnische Mitteilungen (West Germany); 1988 July/August; 32(4): 149- 59.

Eckerson, W. Users can't afford to wait for promise of broadband. Network World; 1989 February 13; 6(6): 21-2.

Eckhardt, G. Information satellites. Incorporated in the global network. Funkschau (West Germany); 1988 September 9; (19): 44-7.

Ehrke, H.-J. et al. Light valve projector for HDTV with 1600 lm light output using a 400 W lamp. Frequenz (West Germany); 1985 September; 39(9): 257-62.

Emmerson, V.P. Delay lines and filters for HDTV. Broadcast Technology (Canada); 1987 October; 13(2): 61-3.

Emmerson, V.P. Delay lines and filters for HDTV. Fernseh und Kino-Technik (West Germany); 1987 November; 41(11): 523-6.

Emmerson, V.P. Delay lines and filters for HDTV. Radio-Television (Czechoslovakia); 1988; 38(4): 40-3.

Ennis, P. Security economics emerges in Washington. Tokyo Business Today (Japan); 1989 June: 60-61.

Eouzan, J.-Y. 1250/50/1 progressive scanning colour camera. Onde Electronique (France); 1989 July/August; 69(4): 43-51.

Eto, Y. et al. An experimental digital VTR for HDTV. SMPTE Journal; 1986 February; 95(2): 215-19.

Eto, Y.; Umemoto, M.; Kawamura, T. Considerations for improvement of an HDTV digital VTR. SMPTE Journal; 1987 February; 2(177- 9).

Evans, R. Electronics: the illusion of recovery. International Management (Great Britain); 1989 May; 44(5): 28-30.

Faroudja, Y.C. NTSC and beyond (TV). IEEE Transactions on

Consumer Electronics; 1988 February; 34(1): 166-78.

Faroudja, Y.; Roizen, J. Improving NTSC to achieve near-RGB performance. SMPTE Journal; 1987 August; 96(8): 750-61.

Faroudja, Y.C.; Roizen, J. A progress report on improved NTSC. Television (Great Britain); 1989 May/June; 26(3): 116-22.

Fehlauer, E. Solution methods for recording of HDTV signals. Fernseh und Kino-Technik (West Germany); 1988 March; 42(3): 105-8.

Feldman, L. Stereo and high definition TV update. DB Sound Engineering Magazine; 1983 October; 17(9): 21-4.

Fenton, B.C. HDTV update. Radio-Electronics; 1988 January; 59(1): 16-17, 66.

Fibush, D.K.; Friedman, J.B. Video pictures of the future; high definition television, television graphics and special effects, video tape formats, microcomputers. Scarsdale, NY: Society of Motion Picture and Television Engineers; 1983.

Fink, D.G. The future of high-definition television: first portion of a report of the SMPTE study group on high-definition television. SMPTE Journal; 1980 February; 89(2): 89-94.

Fink, D.G. The future of high-definition television: conclusion of a report of the SMPTE study group on high-definition television. SMPTE Journal; 1980 March; 89(3): 153-61.

Fisher, D. High definition television: on the brink. Television (Great Britain); 1988 May/June; 25(3): 113-16.

Flaherty, J.A. High-definition television production. SMPTE Journal; 1988 October; 97(10): 844-6.

Flaherty, J.A. Television: the challenge of the future. SMPTE Journal; 1987 September; 96(9): 846-50.

Flannaghan, B. Converting HDTV problems into solutions. International Broadcasting (Great Britain); 1989 April; 12(3): 34-6.

Forrest, J.R. Big screen home TV by satellite. Image Technology (Great Britain); 1988 November; 70(11): 416-18.

Forrest, J.R. Commercial satellite broadcasting for Europe. IEEE Transactions on Broadcasting; 1988 December; 34(4): 443-8.

Forrest, J.R. A route to higher-definition television with DBS. SMPTE Journal; 1987 November; 96(11): 1087-9.

Forrest, J.R. Two crier to world seer: a broadcast odyssey. IEE

Proceedings F, Communication Radar Signal Processing (Great Britain); 1988 February; 135(1): 1-6.

Franken, A.; Rao, N.V. Television camera tubes and solid-state sensors for broadcast applications. SMPTE Journal; 1986 August; 95(8): 799-804.

Freeman, K.G. A multi-standard high-definition television projector. Journal of the Institute of Electronic and Radio Engineering (Great Britain); 1985 February; 55(2): 47-53.

Fresquet, J. CCD technology for imaging. Revista Espana Electronico (Spain); 1988 December; (409): 42-5.

Fujii, A.; Ohtsuki, K. High grade satellite receiver. NEC Technical Journal (Japan); 1989 April; 42(5): 30-6.

Fujiio, T. A short history of HDTV (Hi-Vision). Journal of the Institute of Television Engineers of Japan (Japan); 1988 June; 42(6): 570-8.

Fujimura, R. et al. Hi-vision broadcasting satellite receiving system. NEC Technical Journal (Japan); 1989 April; 42(5): 3-9.

Fujimura, R. et al. Improvement of picture quality on projection TV for high-definition television. NEC Technical Journal (Japan); 1985 August; 38(8): 28-32.

Fujimura, R. et al. MUSE decoder. NEC Technical Journal (Japan); 1988 April; 41(3): 3-10.

Fujino, S.; Kato, H.; Miura, N. An expressway traffic-surveillance system. Mitsubishi Denki Giho (Japan); 1987; 61(4): 65-8.

Fujio, T. Future broadcasting and high-definition TV. NHK Technical Monographs (Japan); 1982 June; (32): 5-101.

Fujio, T. High definition TV. Journal of the Institute of Television Engineers of Japan (Japan); 1980 July; 34(7): 618-20.

Fujio, T. High definition TV. Journal of the Institute of Television Engineers of Japan (Japan); 1981 December; 35(12): 1016-23.

Fujio, T. High definition television systems: desirable standards, signal forms, and transmission systems. IEEE Transactions on Communications; 1981 December; COM-29 (12): 1882-91.

Fujio, T. High resolution TV passes tests, competes for 12 GHZ spectrum. MSN Microwave Systems News; 1982 September; 12(9): 135-48.

Fujio, T. High-definition television systems. Proceedings of IEEE; 1985 April; 73(4): 646-55.

Fujio, T. High-definition wide-screen television system for the future-present. State of the study of HD-TV systems in Japan. IEEE Transactions on Broadcasting; 1980 December; BC- 26(4): 113-24.

Fujio, T. Movement of hi-vision toward the new information-orientated society (HDTV). Journal of the Institute of Electronic and Communication Engineers of Japan (Japan); 1985 December; 68(12): 1290-6.

Fujio, T. Review for high-definition TV. Journal of the Institute of Television Engineers of Japan (Japan); 1985 August; 39(8): 664-8.

Fujio, T. A study of high-definition TV system in the future. IEEE Transactions on Broadcasting; 1978 December; BC-24 (4): 92-100.

Fujio, T. A universal weighted power function of television noise and its application to high-definition TV system design. IEEE Transactions on Broadcasting; 1980 June; BC-26(2): 39- 47.

Fujio, T. A universal weighted power function of television noise and its application to high-definition TV system design. SMPTE Journal; 1980 September; 89(9): 663-9.

Fujio, T. et al. High-definition television system-signal standard and transmission. SMPTE Journal; 1980 August; 89(8): 579-84.

Fujio, T.; Motoki, T.; Sugiura, Y. A study of component coding standards of color television signal for electro-cinematography. NHK Laboratory Note (Japan); 1981 January; (259): 1-11.

Fujita, M. et al. LSIs for improved NTSC receivers. National Technical Reports (Japan); 1988 October; 34(5): 97-104.

Fujita, Y. et al. High-definition television evaluation for remote handling task performance. Transactions of the American Nuclear Society; 1986 November; 53: 486-8.

Fujito, K. et al. PCM optical transmission system for high-definition TV signal-component signal coding type. National

Technical Reports (Japan); 1983 October; 29(5): 650-7.

Fukinuki, T. Signal processing in EDTV and IDTV. Journal of the Institute of Television Engineers of Japan (Japan); 1986 May; 40(5): 357-65.

Fukinuki, T. et al. Fully compatible EDTV for improving both Y and color signals by using a single new subcarrier. IEEE Transactions on Consumer Electronics; 1988 August; 34(3): 469-73.

Gaggioni, H.P. The evolution of video technologies. IEEE Communications Magazine; 1987 November; 25(11): 20-36.

Gainsborough, J. New platinum panaflex. Image Technology (Great Britain); 1986 December; 68(12): 580-81.

Galt, J.; Pantuso, C. Chasing rainbows: a technical overview. SMPTE Journal; 1989 March; 98(3): 179-83.

Garault, T. et al. A digital MAC decoder for the display of a 16/9 aspect ratio picture on a conventional TV receiver. IEEE Transactions on Consumer Electronics; 1988 February; 34(1): 137-46.

Gardiner, P. D-MAC transmission standard. Electronic Technology (Great Britain); 1989 April; 23(4): 64-8.

Garrett, J. 50 years of BBC television - some engineering milestones. EBU Review of Technology (Belgium); 1986 December; (220): 13-25.

Gaspar, J.; Mahler, H.; Gabritsos, G. Comparison of HDTV and film - overall light transfer characteristics. SMPTE Journal; 1989 August; 98(8): 556-562.

Gaus, H.H. A digital HDTV test picture generator for the creation of time-varying patterns. Fernseh und Kino-Technik (West Germany); 1986 January; 40(1): 3-7.

Gee, J. Broadcast bulletin: ITT and Philips see cash in TV chips. Electronic Business; 1988 May 15; 14(10): 108-109.

Gerbasi, G.A. The new HDTV television system. Revista Telegrafica Electronica (Argentina); 1985 May; 73(862): 438; ISSN 42.

Giller, H. TV test instrument with enhancement facilities: (Grundig FG70 colour generator). Elektronikschau (Austria); 1986 September; 62(9): 62-3.

Girgat, R.-R.; Weinerth, H. Key technology; micro-electronics. X. Applications example: television. Elektronik (West Germany); 1989 April 28; 38(9): 78-86.

Glenn, K.G.; Glenn, W.E. From vision science to HDTV: bridging the gap. Information Dispatch; 1987 February; 3(2): 22-5.

Glenn, W.E.; Glenn, K.G. HDTV compatible transmission system. SMPTE Journal; 1987 March; 96(3): 242-6.

Glenn, W.E.; Glenn, K.G. High definition television compatible transmission system. IEEE Transactions on Broadcasting; 1987 December; BC-33(4): 107-15.

Glider, G. IBM-TV? Forbes; 1989 February 20; 143(4): 72-6.

Golding, J.; Hill, T. The 5050 and 5060 HDTV SPGs. Institute of Broadcast Engineers (Great Britain); 1988 September; 19(225): 100-3.

Goldman, T. The consumer electronics picture. Marketing Communications; 1987 December; 12(12): 28-34.

Goodman, D. Video 1990. Radio-Electronics; 1982 January; 53(1): 77-80.

Goodwin, R.I. On the question of a suitable line rate for HDTV. ABU Technical Review (Japan); 1987 March; (109): 3-7.

Gordon, S.L.; Meyers, A.W. INTELSAT goes for the gold in '88. Telephony; 1988 September 26; 215(13): 44-51.

Grabmann, H. TV production: video components in the transmission van. Funkschau (West Germany); 1989 June 2; (12): 56-8.

Gregory, G. The promise of high definition TV. Electronic Australia (Australia); 1985 April; 47(4): 16-21.

Grewlich, K.W. HDTV, point of contention: global tele-presence. Funkschau (West Germany); 1989 July 28; (16): 44-7.

Grewlich, K.W. HDTV: the struggle for telepresence. Transatlantic Data Communication Reports; 1989 April; 12 (4): 17-23.

Grigsby, J. Sizzling scientists. Financial World; 1989 April 18; 158(8): 53-4.

Grimaldi, J.L. High-definition still-picture scanner. Revue Radiodiffusion et Television (France); 1988 October; 22(103): 18-20.

Groenewold, G. Rules of the game. UNIX Review; 1989 July; 7(7): 31-34.

Groenewold, G. Uneasy bedfellows - the FCC and HDTV: standards issues. UNIX Review; 1989 June; 7(6): 38-44.

Grotz, K. et al. A combination of DPCM and motion-compensated frame interpolation for the encoding of 34 Mbit/s colour TV signals. III. ANT Nachrichtentechnische (West Germany); 1988 September; 42(9): 247-51.

Guichardaz, P. Europe's challenge in HDTV. Europe; 1989 January/February; (283): 18-19.

Guillier, M. Television improvements: beyond the technology choices. Bulletin IDATE (France); 1986 November; (25): 166-74.

Haase, H.J. HDTV in the cinema. Sound (Switzerland); 1988 November; (11): 54-6.

Habermann, W. International discussion of HDTV: possible future standards. Fernseh und Kino-Technik (West Germany); 1986 September; 40(9): 410-18.

Habermann, W.; Wood, D. Images of the future. The EBU's part to date in HDTV system standardization. EBU Review of Technology (Belgium); 1986 October; (219): 267-80.

Happe, B. Viewing video (guidelines to the small screen). BKSTS Journal (Great Britain); 1983 June; 65(6): 334-7.

Harada, N. Amorphous-Si-films-laminated CCD-type image pick-up device for HDTV of 2 million picture elements. Nikkei Electronics (Japan); 1988; 441(207): 12.

Harada, S.; Kinpara, A. The trends of satellite broadcasting: the present situations and prospects of satellite broadcasting in Japan. Journal of the Institute of Television Engineers of Japan (Japan); 1987 April; 41(4): 309-12.

Harashima, H. Intelligent image coding and communication. Journal of the Institute of Television Engineers of Japan (Japan); 1988 June; 42(6): 519-25.

Harsdorff, G. TV of the future has high definition screens. NET Nachrichten Elektronik und Telematik (Germany); 1985 May; 39(5): 200-2.

Hashimoto, Y. Experimental HDTV digital VTR with a bit rate of 1 Gbps. IEEE Transactions on Magnetics; 1987 September; MAG-23(5 part 2): 3167-72.

Hashimoto, Y.; Nakaya, H.; Yoshinaka, T. An experimental HDTV digital VTR with a bit rate of 1.188 Gbps. IEEE Transactions on Broadcasting; 1987 December; BC-33(4): 203-9.

Haskell, B.G. High definition television (HDTV) Compatibility and distribution. IEEE Transactions on Communications; 1983

December; COM-31(12): 1308-17.

Haskell, B.G. Semicompatible high definition television using field differential signals. IEEE Transactions on Communications; 1986 October; COM-34(10): 1031-7.

Hatanaka, Y.; Mimura, H.; Aoyama, M. High definition TV imaging devices using amorphous silicon. Bulletin of the Research Institute for Electronics. Shizuoka University (Japan); 1985; 20(1): 91-7.

Hatori, M. Image information theory. Journal of the Institute of Television Engineers of Japan (Japan); 1984 July; 38(7): 634-7.

Hatori, M.; Okoshi, T. Digital television: a prospect of digital television. Journal of the Institute of Television Engineers of Japan (Japan); 1986 March; 40(3): 198-203.

Hausdorfer, M. Towards digital HDTV recording. Fernseh und Kino-Technik (West Germany); 1989; 43(7): 364-7.

Hawker, P. High-definition television. Middle East Electronics (Great Britain); 1983 February; 6(2): 18-20.

Hawley, G. Fiber optics comes full circle. Network World; 1987 January 19; 4(3): 28.

Hayashi, K. Research and development on high-definition television in Japan. SMPTE Journal; 1981 March; 90(3): 178- 86.

Hayashi, K.; Sugimoto, M.; Mitsubashi, T. High definition television. Journal of the Institute of Television Engineers of Japan (Japan); 1984 July; 38(7): 610-14.

Hayashi, Y. High definition TV (HDTV) system. Sharp Technical Journal (Japan); 1988 October; (40): 7-13.

Heaton, J. And in ten years. I&CS (Instrumentation & Control Systems); 1989 May; 62(5): 99-101.

Heeren, H./Talmi, M. Image storing for HDTV. ITG-Fachberichte (West Germany); 1988; 103: 127-32.

Heine, K. The way to HDTV receiving equipment. Critical comments from the point of view of a TV set manufacturer. Funkschau (West Germany); 1987 January 30; (3): 39-41.

Heiss, R.; Rousseau, A.; Verbiest, W. Video coding family for STM and ATM networks. Electrical Communication; 1988; 62 (3-4): 294-301.

Henry, E.W. Advanced television systems. IEEE Transactions on

Broadcasting; 1987 December; BC-33(4): 90-2.

Herbert, R.M. The Baird intermediate film process. Television
(Great Britain); 1987 May/June; 24(3): 134-7.

Higuchi, S. et al. An experimental color-under VCR for IDTV/EDTV
system. IEEE Transactions on Consumer Electronics; 1988
February; 34(1): 228-35.

Hills, R.C. Fifty years of high-definition television
transmission. Journal of the Institute of Electronic and
Radio Engineers (Great Britain); 1986 January; 56(1): 1-15.

Hioki, T. et al. Hi-vision optical video disc. IEEE Transactions
on Consumer Electronics; 1988 February; 34(1): 72-7.

Holoch, G. HDTV standards: development within existing TV
standards. Fernseh und Kino-Technik (West Germany); 1988
April; 42(4): 177-84.

Hopkins, R. Advanced television systems. IEEE Transactions on
Consumer Electronics; 1988 February; 34(1): 1-15.

Hopkins, R. Advanced television systems. IEEE Transactions on
Consumer Electronics; 1986 May; CE-32(2): xi-xvi.

Hopkins, R. Advanced television systems. II. CED; 1988 June;
14(6): 62, 64, 66, 68, 70, 72, 74.

Hsu, S.C. The Kell factor: past and present. SMPTE Journal; 1986
February; 95(2): 206-14.

Hurault, J.-P; Marie, G. The EUREKA 95 programme. Onde
Electronique (France); 1989 July/August; 69(4): 7-12.

Hurault, J.P.; Marie, G. The Eureka-95 programme. Revue
Radiodiffusion et Television (France); 1988 October;
22(103): 1-5.

Hurtado, E.G. Radio and television, at present and within the
next twenty years. Revista Espana Electronico (Spain); 1989
March; 28(316): 14-17.

Ibanez, E.B. Rear projection TV systems. Mundo Electronico
(Spain); 1987 October; (176): 61-7.

Ikeda, S. et al. HDTV production switcher. NEC Technical Journal
(Japan); 1985 May; 38(5): 28-32.

Infante, C. CRTs - present and future. Information Display; 1988
December; 4(12): 8-11.

Iredale, R.J. HD-PRO: a new global high-definition video
production format. SMPTE Journal; 1989 June; 98(6):
439-443.

Iredale, R.J. High definition NTSC broadcast protocol. IEEE
        Transactions on Broadcasting; 1987 December;
        BC-33(4): 161- 9.
Iredale, R.J. Proposal for a new high-definition NTSC broadcast
        protocol. SMPTE Journal; 1987 October; 96(10): 959-70.
Ishibashi, M.; Fujii, T. A 1/2 inch high-definition baseband VCR.
        Mitsubishi Electric Advances (Japan); 1987 December;
        41: 15- 18.
Ishibashi, M.; Fujii, T. A home-use high-definition VCR.
        Mitsubishi Denki Giho (Japan); 1986; 60(3): 28-30.
Ishida, J.; Nishizawa, T.; Kubota, K. High definition television
        broadcasting by satellite. IEEE Transactions on
        Broadcasting; 1982 December; BC-28(4): 165-71.
Ishida, T. et al. A 70-mm film laser telecine for high-definition
        television. SMPTE Journal; 1983 June; 92(6): 629-35.
Ishida, T.; Hirabayashi, H. Laser telecine for high-definition
        television. NHK Laboratory Note (Japan); 1985 December;
        (324): 1-15.
Ishida, T.; Soejima, S.; Ichikawa, Y. Present situation of
        Japanese satellite broadcasting for experimental purpose.
        IEEE Transactions on Broadcasting; 1979 December; BC-25
        (4): 105-12.
Isnardi, M.A. et al. Advanced compatible television: a progress
        report. SMPTE Journal; 1989 July; 98(7): 484-95.
Isnardi, M.A.; Smith, T.R.; Roeder, B.J. Decoding issues in the
        ACTV system. IEEE Transactions on Consumer Electronics;
        1988 February; 34(1): 111-20.
Isnardi, M.A. et al. Encoding for compatibility and
        recoverability in the ACTV system. IEEE Transactions on
        Broadcasting; 1987 December; BC-33(4): 116-23.
Isnardi, M.A. et al. Single channel, NTSC compatible widescreen
        EDTV system. Image Technology (Great Britain); 1988 April;
        70(4): 118-119.
Isono, H. Objective evaluation system of display image sharpness.
        Transactions of the Institute of Electronic Information &
        Communication Engineers. D (Japan); 1987 February;
        J70D(2): 474-81.
Isono, H. A real-time measurement system to evaluate image
        sharpness for CRT display. System Computers of Japan

(Japan); 1988 January; 19(1): 58-67.

Isono, H.; Yasuda, M. Television: 3-D picture without flicker. Funkschau (West Germany); 1989 June 16; (13): 59-62.

Isozaki, Y. et al. 1-inch SATICON for high-definition color television cameras. IEEE Transactions on Electronic Devices; 1981 December; KED-28(12): 1500-7.

Ito, S. Attenuation on an Earth-space path in rain at 22 GHz. NHK Laboratory Note (Japan); 1987 October; (351): 1-9.

Itoga, M. et al. Two channel FM recording for high-definition baseband signals. IEEE Transactions on Consumer Electronics; 1988 February; 34(1): 78-84.

Itoga, M. et al. Wideband recording technology for high-definition baseband VCRs. IEEE Transactions on Consumer Electronics; 1987 August; CE-33(3): 203-9.

Ivall, T. Eureka 95-a world standard? Electronic Wireless World (Great Britain); 1988 September; 94(1631): 845-50.

Iwaoka, A.; Fukuchi, Y.; Miwa, M. 400 MBIT/SEC PCM optical transmission equipment. National Technical Reports (Japan); 1983 October; 29(5): 658-63.

Izumikawa, S. Moving closer to high-definition TV: scanning density improves picture quality. AEU (Japan); 1985 September: 97-100.

Jacobsen, M. Picture enhancement for PAL-coded TV signals by digital processing in TV receivers. SMPTE Journal; 1983 February; 92(2): 164-9.

Jain, G.C. Advances in video technology and their impact. Electronic Information and Planning (India); 1983 November; 11(2): 51-64.

Jares, V. 1 inch Saticon for high-definition TV cameras. Slaboproudy Obzor (Czechoslovakia); 1983 August; 44(8): 398- 403.

Jares, V. A diode electron gun with small divergence of electron beam. Slaboproudy Obzor (Czechoslovakia); 1989 January; 50(1): 39-41.

Jeandon, J.-P.; Miralles, P. High-definition television: revolution or evolution? Bulletin IDATE (France); 1986 November; (25): 648-55.

Johnson, T. Strategies for higher-definition television. London: Ovum; 1983.

Johnson, T. Japanese preparations for HDTV and DBS.
    Nachrichtentechnische Zeitschrift NTZ (Germany); 1983
    June; 36(6): 378-81.
Johnstone, B. Programming better quality TV. Far Eastern Economic
    Review (Hong Kong); 1988 August 11; 141(32): 52-4.
Johnstone, B. Standards of vision. Far Eastern Economic Review
    (Hong Kong); 1989 March 9; 143(10): 77.
Jones, G.E. Will high-definition TV upset CG standards? Computer
    Graphics World; 1985 October; 8(10): 99-102, 104.
Jou, L.; Su, S.F.; Lenart, J. A general purpose optical switching
    concept. Proceedings of SPIE - The International Society for
    Optical Engineering; 1988; 841: 221-4.
Jurgen, R.K. Chasing Japan in the HDTV race. IEEE Spectrum; 1989
    October; 26(10): 26-30.
Jurgen, R.K. Consumer electronics. IEEE Spectrum; 1989 January;
    26(1): 59-61.
Jurgen, R.K. High-definition television update. IEEE Spectrum;
    1988 April; 25(4): 56-62.
Jurgen, R.K. The problems and promises of high-definition
    television (colour TV). IEEE Spectrum; 1983 December;
    20(12): 46-51.
Jurgen, R.K. Technology '88: consumer electronics. IEEE
    Spectrum; 1988 January; 25(1): 56-8.
Kaiser, A.; Mahler, H.W.; McMann, R.H. Resolution requirements
    for HDTV. Television (Great Britain); 1985 April; 22(2):
    68- 72.
Kaiser, A.; Mahler, H.W.; McMann, R.H. Resolution requirements
    for HDTV based upon the performance of 35 mm
motion-picture
    films for theatrical viewing. SMPTE Journal; 1985 June;
    94(6): 654-9.
Kakimori, N. et al. Automatic purity measurement system for color
    TV set. Sharp Technical Journal (Japan); 1986; (36): 57-64.
Kalish, D. Creative concepts: picture this. Marketing & Media
    Decisions; 1989 March; 24(3): 32-33.
Kanazawa, M. A rear-projection display for high-definition
    television. NHK Laboratory Note (Japan); 1986 January;
    (325): 1-11.
Kanazawa, M.; Mitsuhashi, T. A 50-inch diagonal rear-projection

display. NHK Laboratory Note (Japan); 1988 March; (356): 1- 11.

Kasahara, Y.; Inaba, R. Temperature-stabilized wide-band ultrasonic delay line. Review of Scientific Instrumentation; 1981 March; 52(3): 443-6.

Katoh, N. et al. Handy color camera for high definition television system. Sharp Technical Journal (Japan); 1986; (34): 109-14.

Kawai, K. et al. IDTV receiver. IEEE Transactions on Consumer Electronics; 1987 August; CE-33(3): 181-91.

Kawamura, T. et al. A new high-resolution pickup tube for live X-ray topography. NHK Laboratory Notes (Japan); 1984 December; (309): 1-12.

Kawamura, T.; Tanada, J. HDTV (Hi-Vision). V. TV cameras. Journal of the Institute of Television Engineers of Japan (Japan); 1988 October; 42(10): 1120-7.

Kawamuta, T. Improvement of camera tube sensitivity. Journal of the Institute of Television Engineers of Japan (Japan); 1988 August; 42(8): 780-6.

Kawashima, M.; Nishida, T.; Yamamoto, K. Projection displays for Hi-Vision (HDTV). National Technical Reports (Japan); 1988 October; 34(5): 66-74.

Kawashima, M.; Yamamoto, K.; Kawashima, K. Display and projection devices for HDTV. IEEE Transactions on Consumer Electronics; 1988 February; 34(1): 100-10.

Kay, R. Large screen projection for professional uses. Kep-es Hangtechnika (Hungary); 1987 December; 33(6): 185-9.

Kays, R. Large screen projection for professional applications. Fernseh und Kino-Technik (West Germany); 1987 April; 41(4): 125-30.

Kays, R. Large screen projection for professional applications. Image Technology (Great Britain); 1987 May; 69(5): 155-58.

Kecskes, P. High definition television chain. Kep-es Hangtechnika (Hungary); 1986 December; 32(6): 189-91.

Keirstead, P. Desperate race to save US terrestrial TV. International Broadcasting (Great Britain); 1989 June; 12(5): 46, 48, 50, 52-3.

Keirstead, P. Dragging consumers into the 21st century (HDTV standards). International Broadcasting (Great Britain); 1988

December; 11(10): 42, 44-5.

Keirstead, P. Who will define high definition? (television).
International Broadcasting Systems Operations; 1988 March;
11(2): 10-12.

Kennedy, M.C. The global standards dilemma-agreement or anarchy?
III. Television (Great Britain); 1989 March/April; 26(2):
68-72.

Khesin, A.Ya.; Shteinberg, A.L. A high definition television
system. Tekhnika Kino i Televideniya (USSR); 1985;
(9): 64- 6.

Khleborodov, V.A. On choosing a single world HDTV standard.
Tekhnika Kino i Televideniya (USSR); 1986 November;
(11): 49-51.

Khleborodov, V.A. Towards a single worldwide HDTV standard.
Tekhnika Kino i Televideniya (USSR); 1988 February;
(2): 37-8.

Kido, T.; Kondo, T. Putting HDTV to practical use has become
realistic target. JEE (Japan); 1986 March; 23(231): 30-5.

Kido, T.; Sugimoto, A.; Morita, S. Video production equipment for
HDTV. Journal of the Institute of Television Engineers of
Japan (Japan); 1985 August; 39(8): 695-9.

Kim, G. Getting better S/N and better pictures. CED; 1988 March;
14(3): 20, 22, 24.

Kimura, E.; Ninomiya, Y. A high-definition satellite television
broadcast system-'MUSE'. Journal of the Institute of
Electronic & Radio Engineers (Great Britain); 1985 October;
55(10): 353-6.

Kimura, H.; Nakagawa, K. F-1. 6G system overview. Review of the
Electrical Communication Laboratories (Japan); 1987 May;
35(3): 219-225.

Kindel, S. Pictures at an exhibition. Forbes; 1983 August 1;
132(3): 137-139.

Kindel, S. A sharper image. Financial World; 1988 August 23;
157(18): 35-6.

Kira, K.; Kashigi, K. HDTV. 8. Video equipment. Journal of the
Institute for Television Engineers of Japan (Japan); 1989
January; 43(1): 39-45.

Kirby, R.C. Broadcasting and international standards. SMPTE
Journal; 1988 September; 97(9): 720-2.

Kishimoto, R.; Sakurai, N,; Ishikura, A. Bit-rate reduction in
    the transmission of high-definition television signals.
    SMPTE Journal; 1987 February; 96(2): 191-7.
Kishimoto, R.; Yoshino, K.; Ikeda, M. Fiber-optic digital video
    distribution system for using high-definition television
    signals using laser-diode optical switch. IEEE Journal of
    Select Areas of Communication; 1988 August; 6(7): 1079-86.
Kiuchi, Y. Pickup techniques. Journal of the Institute of
    Television Engineers of Japan (Japan); 1986 July; 40(7):
    668-72.
Klaas, L.; Hofker, U.; Reuter, T. A procedure for the sampling-
    rate reduction of the colour difference signals in a future
    digital HDTV-system. A E Ue (Germany); 1985 May-June;
    39(3): 161-6.
Klemmer, W. Camera technology for HDTV. Fernseh und
    Kino-Technik (West Germany); 1985 May; 39(5): 224-8.
Klemmer, W. KCH 1000-a multistandard HDTV camera system.
    Fernseh und Kino-Technik (West Germany); 1988 May; 42
    (5): 215-20.
Klemmer, W. Multi-standard HDTV camera. Electronic Wireless
    World (Great Britain); 1988 July; 94(1629): 708-11.
Klemmer, W. The two-dimensional resolution of pickup tubes in
    HDTV camera systems. Image Technology (Great Britain);
    1987 July; 69(7): 328-32.
Klemmer, W. Two-dimensional definition of camera tubes in a high-
    definition television system. Rundfuntechnische Mitteilungen
    (West Germany); 1986; 30(6): 275-80.
Klemmer, W.; Reimers, U. Colour TV camera for high definition TV.
    Bosch Technische Berichte (West Germany); 1985; 7(6):
    254-60.
Kline, D.D. Can Hollywood and HDTV be friends? IEEE
    Transactions on Consumer Electronics; 1988 February;
    34(1): 48-53.
Knapp, K.H. High definition television. Compatible solution
    proposals. Funkschau (West Germany); 1982 September 17;
    (19): 58-60.
Komiyama, S. Subjective evaluation of angular displacement
    between picture and sound directions for HDTV sound
    systems. Journal of the Audio Engineering Society; 1989

April; 37(4): 210-14.

Komiyama, S. Visual factors in sound localization for HDTV. Journal of the Acoustical Society of Japan (Japan); 1987 September; 43(9): 664-9.

Konishi, Y.; Fukuoka, Y. Satellite receiver technologies. IEEE Transactions on Broadcasting; 1988 December; 34(4): 449-56.

Kopitz, D. The role of the EBU in the development and planning of the new broadcasting services. Revue Radiodiffusion et Television (France); 1987 September/October; 21(99): 19-20.

Kotelnikov, A.V.; Falluh, N. The particulars of color rendition with HDTV signal down-conversion. Tekhnika Kino i Televideniya (USSR); 1989 January; (1): 16-18.

Koziol, R. 60 years of cine film form Mortsel. The contributions of Agfa-Gevaert to film and television. Fernseh und Kino-Technik (West Germany); 1986 January; 40(1): 15-18, 21.

Kramer, D. High definition television - HDTV. Bulletin de l'Association Suisse des Electriciens (Switzerland); 1987 September 5; 78(17): 1055-62.

Kriss, M.A.; Liang, J. Today's photographic imaging technology for tomorrow's HDTV system. SMPTE Journal; 1983 August; 92(8): 804-18.

Kriz, M.E. Looking sharp: high-definition television, a new technology that Japan might soon start selling to American consumers, poses high-stakes challenges across the TV spectrum. National Journal; 1988 July 16; 20: 1860-3.

Kubo, T. Development of high-definition TV displays. IEEE Transactions on Broadcasting; 1982 June; BC-28(2): 51-64.

Kubota, K.; Nishizawa, T. A three dimensional noise weighting function and its application to HDTV transmission. Transactions of the Institute of Electronics and Communications Engineers of Japan Part B (Japan); 1986 May; J69B(5): 503-11.

Kuchenbecker, H.-P.; Zimmer, G. Light wave guides-gigantic bandwidth on 0.3 micrometer. Funkschau (West Germany); 1988 October 7; (21): 54-7.

Kuge, T.; Shibaya, H. A flexible method of time base error detection by microprogram control and its application to the high definition TV signal processing. Transactions of the

Institute of Electronics and Communications Engineers of Japan Section E (Japan); 1986 April; E69(4): 501-4.

Kumada, J.; Mitsuhashi, T. Optical fiber transmission for high-definition TV signal. NHK Laboratory Note (Japan); 1978 October; (228): 1-11.

Kummerow, T. Aspects of digital HDTV systems. Frequenz (West Germany); 1983 November/December; 37(11-12): 278-85.

Kummerow, T. Picture and sound transmission in glass fibre broadband local networks. I. Draft plan of system. Nachrichtentechnische Zeitschrift NTZ (Germany); 1985 March; 38(3): 140-4.

Kurashige, M. Self-sharpening effect on resolution in camera tubes. Journal of the Institute of Television Engineers of Japan (Japan); 1983 May; 37(5): 401-7.

Kurashige, M.; Egami, N. Resolution dependence of Saticon for Hi-Vision on signal storage time. Journal of the Institute of Television Engineers of Japan (Japan); 1988 September; 42(9): 992-4.

Kurashige, M. et al. 1-inch magnetic-focus electrostatic-deflection compact Saticon for HDTV. Journal of the Institute of Television Engineers of Japan (Japan); 1985 September; 39(9): 813-21.

Kurashige, M. et al. 1-inch magnetic-focus electrostatic-deflection compact Saticon for HDTV. NHK Laboratory Note (Japan); 1985 November; (322): 1-16.

Kurashige, M. et al. 2/3-inch MS (magnetic-focus electrostatic-deflection) type Saticon for HDTV hand-held color cameras. IEEE Transactions on Consumer Electronics; 1987 February; CE-33(1): 39-46.

Kurashige, M. et al. Super-sensitive HDTV camera tube with the newly developed HARP target. SMPTE Journal; 1988 July; 97(7): 538-45.

Kurashige, M.; Ohmura, T. TV cameras (for HDTV). Journal of the Institute of Television Engineers of Japan (Japan); 1985 August; 39(8): 669-73.

Kurita, T. et al. A practical IDTV system improving picture quality for nonstandard TV signals. IEEE Transactions on Consumer Electronics; 1988 August; 34(3): 387-96.

Kuroda, H. High-frequency ultrasonic delay line for high-definition TV cameras. National Technical Reports (Japan); 1985 February; 31(1): 131-9.

Kurozumi, K. et al. Sound system suitable for high definition television. Journal of the Institute of Television Engineers of Japan (Japan); 1988 June; 42(6): 579-87.

Kusaka, H.; Komoto, T.; Nishizawa, T. Present state of the research on high definition TV. Journal of the Institute of Television Engineers of Japan (Japan); 1976 December; 30(12): 948-56.

Kuwahara, H. Coherent lightwave technology promises to expand optical fiber networks. JEE, Journal of Electronic Engineering (Japan); 1988 November; 25(263): 80-84.

Landau, S. Toward a global high-definition T.V. production standard. Department of State Bulletin. 89; 1989 June; (2147): 48-51.

Lang, H. The criteria for choosing a standard for HDTV colour signals. Fernseh und Kino-Technik (West Germany); 1989; 43(3): 125-30.

Lang, H. View-points in selection of a HDTV colour signal standard. II. Fernseh und Kino-Technik (West Germany); 1989; 43(4): 183-6, 188-90, 193.

Lawcewicz, T.; Wang, R. Video and graphics-the twain shall meet. ESD, Electronics System Design Magazine; 1988 April; 18 (4): 55-7.

Lechner, B.J. Higher resolution, fewer artifacts, TV technology goals. Information Dispatch; 1985 December; 1(12): 12-15.

Leuthold, P.E. Telecommunication in the future. Sysdata (Switzerland); 1989 April 3; 20(4): 29-32.

Lewis, G. High-definition television. Electronic Wireless World (Great Britain); 1988 March; 94(1625): 226-9.

Li, H.; Yu, S. HVBL system and 'compatible transition' approach to the development of the coming generation TV. Tianjin Daxue Xuebao/Journal of Tianjin University; 1988; (3): 33- 40.

Liebsch, W. Architecture and circuit technology of a two-dimensional DCT for video signal coding. ITG-Fachberichte (West Germany); 1988; 103: 135-9.

Lippman, A.B. et al. Single-channel backward-compatible EDTV systems. SMPTE Journal; 1989 January; 98(1): 14-19.

Liu, H.; Aoki, K. Themes for high-resolution full-color display. Transactions of the Institute of Electronics, Information and Communication Engineers, Section E (Japan); 1987 November; E70(11): 1041-43.

LoCicero, J.L.; Pazarci, M.; Rzeszewski, T.S. A compatible high-definition television system (SLSC) with chrominance and aspect ratio improvements. SMPTE Journal; 1985 May; 94(5, part 1): 546-58.

LoCicero, J.L.; Pazarci, M.; Rzeszewski, T.S. Image reconstruction in a wide aspect ratio HDTV system. IEEE Transactions on Communications; 1986 September; COM-34 (9):946-52.

Lodge, J.A. Thorn EMI Central Research Laboratories - an anecdotal history. Physics in Technology; 1987 November; 18(6): 258-68.

Long, T.J. Why non-compatible high-definition television? IBA Technical Review (Great Britain); 1983 November; (21):1-12.

Long, T.; Stenger, L. The broadcasting of HDTV programmes. EBU Review of Technology (Belgium); 1986 October; (219): 297- 314.

Lucas, K.; van Rassell, B. HDB-MAC: a new proposal for high definition TV transmission. IEEE Transactions on Broadcasting; 1987 December; BC-33(4): 170-83.

Lueder, R. Digital video processing in consumer products-today and tomorrow. Nachrichtentechnische Zeitschrift (West Germany); 1988 September; 10(9): 259-65.

Lundgren, C.W.; Venkatesan, P.S. Applications of video on fiber cable. IEEE Communication Magazine; 1986 May; 24 (5)33-49.

Lyner, A.G. Design of HDTV systems: transmission channel noise. IEE Colloquium on 'Noise in Images '; 1987; 17(3): 1-5.

MacDonald, R.I.; Lam, D.K.W. Optoelectronic switch matrices: recent developments. Optical Engineering; 1985 March/April; 24(2): 220-4.

Mackall, M.J. Likely costs of consumer advanced television (ATV) technology. IEEE Transactions on Consumer Electronics; 1989 May; 35(2): 63-71.

Mackay, W.E.; Davenport, G. Virtual video editing in interactive multimedia applications. Communications of the ACM; 1989

July; 32(7): 802-10.

MacKellar, J.C. A review of television receiver design trends. Television (Great Britain); 1984 November/December; 21(6): 315-24.

Maeda, M.; Oyamada, K.; Utsumi, Y. FM-FDM optical transmission for HDTV in broadcasting station. Transactions of the Institute of Electronic and Communications Engineers (Japan) Part B; 1986 September; J69B(9): 904-13.

Mahler, G. Feasibility of light-valve projection for large screen high-definition television. Frequenz (West Germany); 1983 November/December; 37(11-12): 300-6.

Mahler, G. High definition light-valve projection. IEEE Transactions on Consumer Electronics; 1984 November; CE-30(4): 563-7.

Makino. S. Development of high-definition TV systems in present-day Japan. JEE (Japan); 1987 March; 24(243): 28-30.

Maksakov, A.A.; Sorokina, T.G. TV pick-up cameras for HDTV systems. Tekhnika Kino i Televideniya (USSR); 1988 December; (12): 24-9.

Maksakov, A.A.; Sorokina, T.G. The use of a synchronous wobble method for enhancing TV picture sharpness. Tekhnika Kino i Televideniya (USSR); 1988 October; (10): 17-23.

Maksakov, A.A.; Sorokina, T.G. Ways to designing a compatible HDTV broadcasting system. Tekhnika Kino i Televidenya; 1987 October; (10): 21-5.

Malek, J.; Pazderak, J. The possibilities of enhancing conventional TV systems. Slaboproudy Obzor (Czechoslovakia); 1987 February; 48(2): 85-9.

Maltz, M. A new age of videoconferencing. Telephony; 1989 June 26; 216(26): 30-2.

Mann, T. High definition television as it stands today. BKSTS Journal (Great Britain); 1983 September; 65(9): 474-81.

Marbach, W.D.; Port, O. Super television; high-definition TV is rallying a digital revolution. Business Week; 1989 January 30; (3089 (Industrial/Technology Edition)): 56-66.

Marjenkow, A. The use of predistortion for raising the transmission quality of TV and HDTV signals on light wave conductors. Nachrichtentechnische Technik Electronik (East Germany); 1988; 38(11): 406-7.

Markitalo, O. Hopes good for HDTV prelude for the 'nineties. Tele (Swedish Edition) (Sweden); 1988; 1: 15-17.

Marshall, P.G. Do antitrust laws limit U.S. competitiveness? For a hundred years Congress has prohibited American businesses from creating monopolies, fixing prices or otherwise restraining trade. Editorial Research Reports; 1989 July 7: 366-79.

Marshall, P.G. A high-tech, high stakes HDTV gamble. Congressional Quarterly Editorial Research Reports; 1989 February 17: 90-102.

Marshall, P.G. A high-tech, high-stakes HDTV gamble. Editorial Research Reports; 1989 February 17: 90-103.

Martin, E.R. Direct broadcasting satellite system: system characteristics. COMSAT Technical Review; 1981 Fall; 11 (2): 215-26.

Martin, E.R. HDTV-a DS perspective. IEEE Journal of Selective Areas of Communication; 1985 January; SAC-3(1): 76-86.

Martini, H. Large scale projection: HDTV with light valve. Funkschau (West Germany); 1987 August; (17): 42-6.

Martini, H. Light-valve projection for large-screen high-definition TV. I. Nachrichtentechnische Zeitschrift (West Germany); 1987 August; 9(8): 193-203.

Martini, H. Light-valve projection for large screen high-definition television. II. Nachrichtentechnische Zeitschrift (West Germany); 1987 September; 9(9): 237-49.

Martini, H. Light-valve projection for large screen high-definition television. III. Nachrichtentechnische Zeitschrift (West Germany); October 1987; 9(10): 269-76.

Mary, L.; Laporte, A. Data signal processing becomes pervasive. Canadian Electronics Engineering; 1989 April; 33(4): 17-18, 21, 23.

Mathias, H. Gamma and dynamic range needs for an HDTV electronic cinematography system. Image Technology (Great Britain); 1988 March; 70(3): 81-6.

Mathias, H. Gamma and dynamic range needs for an HDTV electronic cinematography system. SMPTE Journal; 1987 September; 96(9): 840-5.

Matsumoto, S. et al. 120/140 Mb/s digital coding system with noise shaping filter for HDTV signals with the aim of developing compact hardware. Journal of the Institute of Television Engineers of Japan (Japan); 1988 December; 42(12): 1372-9.

Matsumoto, S.; Saitio, M.; Murakami, H. 120/140 Mbit/s Compact HDTV codec. KDD Technical Journal (Japan); 1988 October; (138): 1-11.

Matsunaga, T. et al. Space-division optical switching system using laser diode gate matrix switches. Electrical Communications Laboratories Review (Denki Tsushin Kenkyusho Kenkyu Jitsuyoka Hokoku) (Japan); 1989; 38(2): 115-123.

Matsushima, K. Color blooms in high-performance printer using one-pass thermal transfer process. OEP Office Equipment & Products (Japan); 1984 October; 13(72): 66-9.

Matsushita, M.; Yokoyama, S. Experience on operating a DBS system (BS-2) in Japan. IEEE Transactions on Broadcasting; 1988 December; 34(4): 430-4.

Matsuzawa, A. A/D, D/A converter. Journal of the Institute of Television Engineers of Japan (Japan); 1986 December; 40(12): 1176-80.

Matsuzawa, A. et al. Ultra-high-speed 8-bit A/D converter. National Technical Reports (Japan); 1986 February; 32(1): 99-105.

Matsuzawa, A. et al. A video-rate 8 bit parallel A/D converter using trench isolation. Transactions of the Institute of Electronic Information and Communication Engineers (Japan); 1987 April; E70(4): 230-2.

McCrirrick, T.B. Colour television: yesterday, today and tomorrow. IEE Proceedings, Part A: Physical Science, Measurement and Instrumentation, Management and Education, Reviews; 1989 January; 136(1): 1-7.

Meeker, G.W. High definition and high frame rate compatible NTSC broadcast television system. IEEE Transactions on Broadcasting; 1988 September; 34(3): 313-22.

Meieran, H. Looking beyond the robot hype. Nuclear Engineering International; 1989 March; 34(415): 41-43.

Melwig, B. Digital or higher definition. Revue Radiodiffusion et television (France); 1986 November/December; 20(95):10-12.

Melwig, R. Characterisation of television on pictures. Revue

Radiodiffusion et Television (France); 1985 June/August; 19(88): 16-18.

Melwig, R. Colorimetry in HDTV: up-to-date solutions for a new system. Proceedings of SPIE International Society for Optical Engineering; 1986 594; (41-8).

Melwig, R. NTSC-compatible approaches to HDTV. EBU Technical Review (Belgium); 1988 December; (232): 248-259.

Melwig, R. Towards high-definition television. Echo des Recherches (France); 1986; (126): 9-20.

Melwig, R.; Nasse, D. Introducing HDTV at the time of digital TV: yes, but in what way? Bulletin IDATE (France); 1986 November; (25): 155-65.

Mendrala, J.A. Electronic cinematography for motion-picture film. SMPTE Journal; 1987 November; 96(11): 1090-4.

Mertens, H. The future evolution of broadcasting standards. Television (Great Britain); 1985 February; 22(1): 4-12.

Messerschmid, U. Technical possibilities for improving picture and sound. PTT Technische Mitteilungen (Switzerland); 1983; 61(12): 448-50.

Messerschmid, U. Television standards for communication satellites--the D2-MAC/packet methods in the MAC system family. Fernseh und Kino-Technik (West Germany); 1985 October; 39(10): 472-80.

Metsugi, Y.; Togawa, M.; Sugimoto, T. Cathode ray tube for video projector. NEC Technical Journal (Japan); 1985 August; 38(8): 125-7.

Mica, G. TV satellites: HDTV from orbit. Funkschau (West Germany); 1989 May 19; (11): 40-6.

Micic, L. Television techniques-free programmed. Funkschau (West Germany); 1988 October 21; (22): 46-7 supplement.

Miki, T.; Ishida, J. Transmission technologies on HDTV. Journal of the Institute of Electronic & Communication Engineers of Japan (Japan); 1985 June; 68(6): 635-42.

Mikkela, O. Colorimetric problems in high-definition television. EBU Technical Review (Belgium); 1988 April; (228): 60-67.

Minaguchi, H.; Nomura, T. An overview of recent technology of digital video recording. Journal of the Institute of Television Engineers of Japan (Japan); 1987 December; 41(12): 1153-61.

Minomiya, Y. HDTV (Hi-Vision). III Muse system. Journal of the
    Institute of Television Engineers of Japan (Japan); 1988
    August; 42(8): 823-30.
Mitsuda, K.; Kashihara, S.; Suzuki, Y. 40-inch color CRT for
    high-definition TV. National Technical Reports (Japan); 1987
    April; 33(2): 176-83.
Mitsuhashi, T. A 1125-scan-line wideband contour corrector. NHK
    Technical Journal (Japan); 1976; 28(2): 76-88.
Mitsuhashi, T. HDTV. II. Fundamental parameters and standards of
    HDTV (Hi-Vision). Journal of the Institute of Television
    Engineers of Japan (Japan); 1988 July; 42(7): 742-50.
Mitsuhashi, T. A study of the relationship between scanning
    specifications and picture quality (TV receivers). NHK
    Laboratory Note (Japan); 1980 October; (256): 1-9.
Miyahara, M. Perception of correlation in spatial-temporal image
    fields and physical factors related to HDTV picture quality-
    assessment words; natural, smooth, continuous and stable.
    Journal of the Institute of Television Engineers of Japan
    (Japan); 1986; 40(1): 46-53.
Miyahara, M. Quality assessments for visual service. IEEE
    Communications Magazine; 1988 October; 26(10): 51-60, 81.
Miyahara, M. et al. Physical factors related to the HDTV picture
    quality-brilliance and gloss. Journal of the Institute of
    Television Engineers of Japan (Japan); 1986; 40(11): 1106-12.
Mochizuki, K. et al. Scan-converter for high quality picture TV
    system. NEC Technical Journal (Japan); 1988 April;
    41(3): 24-9.
Moehrmann, K.H. Coding of video signals for digital transmission.
    Telecom Report; 1987 September/October; 10(5): 266-270.
Moffat, B. Television engineering research in the BBC, today and
    tomorrow. SMPTE Journal; 1988 January; 97(1): 17-24.
Mohrmann, K.H. Coding of video signals for digital transmission.
    Telcom Report (West Germany); 1987 September/October;
    10(5): 266-71.
Mokhoff, N. A step toward 'perfect' resolution (high-definition
    TV). IEEE Spectrum; 1981 July; 18(7): 56-8.
Monozawa, H. The state of HDTV development in Japan. AEU
    (Japan); 1989; (3): 44-9.
Morgan, W.L. TV: searching for a standard. Satellite

Communications; 1986 May; 10(5): 38-9.

Morishita, M. et al. MUSE system high definition television receiver. NEC Technical Journal (Japan); 1985 May; 38(5): 41-6.

Moriya, R. Advances in magnetic and optical recording-digital video tape recorder. Journal of the Institute of Television Engineers of Japan (Japan); 1988 April; 42(4): 338-46.

Morris, J.S. HDTV subjective tests. Annual Review of the Phillips Research Laboratory; 1986: 56-59.

Mothersole, P.L. Developments in broadcasting technology and their effect on TV receiver design. Electronics and Power (Great Britain); 1986 November/December; 32(11): 791-5.

Mothersole, P.L. Evolution of the domestic receiver - future trends. Television (Great Britain); 1987 May/June; 24(3): 147-50, 152-3, 155.

Motoki, N.; Makino, S. HDTV. Journal of the Institute of Television Engineers of Japan (Japan); 1988 July; 42(7): 655-8.

Muller-Romer, F. Better television for more money? High-definition television. Funkschau (West Germany); 1982 September 17; (19): 50-2.

Muller-Romer, F. Consequences for the TV technology using further developments of electronics. Fernseh und Kino-Technik (West Germany); 1981 October; 35(10): 365-70.

Muller-Romer, F. Future television system. Fernseh und Kino-Technik (West Germany); 1989; 43(6): 286-92, 94.

Muller-Romer, F. High definition television (HDTV)-research, development and experiments in a Federal Republic (BRD). Fernseh und Kino-Technik (West Germany); 1987 June; 41 (6): 233-6.

Muller-Romer, F. Improvement possibilities for TV pictures and sound. 1. Television with higher picture definition. Fernseh und Kino-Technik (West Germany); 1983 January; 37(1): 9-14.

Muller-Romer, F. Panoramic screen for television. Standards discussion. Funkschau (West Germany); 1989 April 7; (8): 54- 6.

Murakami, H. et al. Planar pulse discharge panel with internal memory for a colour TV display. Journal of the Institute of

Television Engineers of Japan (Japan); 1986 October; 40(10): 953-60.

Murakami, H.; Hashimoto, H.; Hatori, Y. Quality of band-compressed TV services. IEEE Communications Magazine; 1988 October; 26(10): 61-9, 81.

Murakami, H.; Kaneko, R.; Sega, S. A pulse discharge panel with internal memory for a color TV display. NHK Technical Journal (Japan); 1986; 38(2): 55-69.

Murakami, H.; Katoh, T. Planar pulse discharge panel for a TV display. IEEE Electron Device Letters; 1985 March; EDL-6 (3): 132-4.

Murata, T. et al. Practical TV ghost canceller using 2-stage transversal filter. IEEE Transactions on Consumer Electronics; 1983; CE-29(3): 350-7.

Nadan, J.S. A glimpse into future television. BYTE; 1985 January; 10(1): 135-50.

Nagata, S. Fusional characteristics of binocular parallax as a function of viewing angle and viewing distance of stereoscopic picture. Journal of the Institute of Television Engineers of Japan (Japan); 1989 March; 43(3): 276-81.

Naimpally, S. et al. Integrated digital IDTV receiver with features. IEEE Transactions on Consumer Electronics; 1988 August; 34(3): 410-19.

Nakagawa, I. Frequency analysis of MUSE signal. Journal of the Institute of Television Engineers of Japan (Japan); 1986; 40(11): 1126-32.

Nakagawa, I. et al. Development of HDTV receiving equipment based on band compression technique (MUSE). IEEE Transactions on Consumer Electronics; 1986 November; CE-32(4): 759-68.

Nakagawa, M. Satellite broadcasting in Japan. AEU (Japan); 1989; (3): 50-3.

Nakajima, K.; Murakami, M. Color monitor for HDTV. JEE, Journal of Electronic Engineering (Japan); 1988 March; 25(255): 82-4.

Nakamura, M. et al. A 10-bit 65 MHz glitch-free video D/A converter. IEEE Transactons on Consumer Electronics; 1985 August; CE-31(3): 592-600.

Nakanishi, T.; Yamauchi, H.; Yoshimura, H. CMOS geometrical mapping processor GMP-1. Transactions of the Institute for

Electronic Information Communications Engineers. C-II (Japan); 1989 May; J72C-II(5): 354-61.

Nasse, D.; Chatel, J. Toward a world studio standard for high-definition television. SMPTE Journal; 1989 June; 98(6): 434-38.

Neff, R.; Magnusson, P. Rethinking Japan - the new harder line toward Tokyo. Business Week; 1989 August 7; (3118 (Industrial/Technology Edition)): 44-52.

Negroponte, N.; Papert, S.; Wiesner, J.B. MIT's media center. Communicator's Journal; 1984 July/August/September/October; 2(4): 36-43.

Nguyen, T.T. High definition television. Tekhnika Kino i Televideniya (USSR); 1985; (11): 26-32.

Nickelson, R.L. Technical bases for the broadcasting-related work of the ORB(2) Conference-a summary of the CCIR intersessional work. Telecommunications Journal (English Edition) (Switzerland); 1988 August; 55(8): 548-59.

Niitsu, S.; Okada, Y. Improved definition TV system LSI. NEC Technical Journal (Japan); 1987 October; 40(10): 187-91.

Ninomiya, Y. An accurate 8-bit A/D converter sampling at 100 MHZ. IEEE Transactions on Communications; 1981 September; COM- 29(9): 1353-7.

Ninomiya, Y. Bandwidth compression for high definition TV. Journal of the Institute of Television Engineers of Japan (Japan); 1983 November; 38(11): 960-7.

Ninomiya, Y. Coding systems for HDTV. Journal of the Institute of Electronic Information Communication Engineers (Japan); 1988 July; 71(7): 676-82.

Ninomiya, Y. An A/D converter for high-definition TV. Journal of the Institute of Television Engineers of Japan (Japan); 1981 February; 35(2): 120-35.

Ninomiya, Y. Fabrication of a parabolic mirror by a rotational method for a high-definition television display. NHK Laboratory Note (Japan); 1980 January; (245): 1-12.

Ninomiya, Y. High definition television. Journal of the Institute of Television Engineers of Japan (Japan); 1985 January; 39(1): 66-7.

Ninomiya, Y. Satellite-applied HDTV broadcasting transmission system MUSE. Nikkei Electronics (Japan); 1987; (433):

189- 212.

Ninomiya, Y. Transmission equipment (for HDTV). Journal of the Institute of Television Engineers of Japan (Japan); 1985 August; 39(8): 680-4.

Ninomiya, Y. et al. Concept of the MUSE system and its protocol (HDTV). NHK Laboratory Note (Japan); 1987 July; (348): 1-34.

Ninomiya, Y. et al. HDTV broadcast system via broadcasting satellite-MUSE. Journal of the Institute of Television Engineers of Japan (Japan); 1988 May; 42(5): 468-77.

Ninomiya, Y. et al. A HDTV broadcasting system utilizing a bandwidth compression technique-MUSE. IEEE Transactions on Broadcasting; 1987 December; BC-33(4): 130-60.

Ninomiya, Y.; Ohtsuka, Y.; Izumi, Y. NHK proposes high-definition TV using MUSE bandwith compression. JEE (Japan); 1985 March; 22(219): 40-4.

Ninomiya, Y.; Ohtsuka, Y.; Izumi, Y. A single channel transmission system for HD-TV satellite broadcasting: MUSE.Transactions of the Institute of Electronic & Communications Engineers (Japan) part 1; 1985 April; J68D (4): 647-54.

Ninomiya, Y.; Ohtsuka, Y.; Izumi, Y. A single channel HDTV broadcast system-the MUSE. NHK Laboratory Notes (Japan); 1984 September; (304): 1-12.

Nishida, Y. et al. A new architecture for solid-state image sensors. Journal of the Institute of Television Engineers of Japan (Japan); 1987 November; 41(11): 1061-7.

Nishida, Y. et al. Wide dynamic range HDTV image sensor with aliasing supression. IEEE Transactions on Consumer Electronics; 1988 August; 34(3): 506-12.

Nishihata, M.; Mukumoto, M. High-definition television transmission system. Journal of the Institute of Television Engineers of Japan (Japan); 1985 July; 39(7): 609-12.

Nishino, K. et al. A 37" color TV for extended-definition reception. Mitsubishi Denki Giho (Japan); 1989; 63(3): 25-8.

Nishizawa, T. Some examples of EDTV and IDTV systems. Journal of the Institute of Television Engineering of Japan (Japan); 1986 May; 40(5): 366-75.

Nobutoki, S.; Nonaka, Y. Technical progress report on the SATICON high-performance camera tube. Hitachi Review (Japan); 1983 June; 32(3): 121-4.

Nonomiya, Y. et al. Development for the MUSE system. NHK Technical Journal (Japan); 1987; 39(2): 76-111.

Nosov, O.G. An experimental HDTV VTR. Tekhnika Kino i Televideniya (USSR); 1989 May; (5): 65-6.

Novakovskaya, O.S. Some problems of reproducing high definition TV pictures. Tekhnika Kino i Televideniya (USSR); 1985 July; (7): 34-8.

Novakovskii, S.V. Further ways and forms of TV broadcast progress. Tekhnika Kino i Televideniya (USSR); 1983; (11): 37-40.

Novakovsky, S.V. et al. The main problems of developing higher definition television. Tekhnika Kino i Televideniya (USSR); 1986; (1): 21-4.

Novakovskii, S.V.; Kotelnikov, A.V.; Falluh, N. On the colorimetry of new TV systems. Tekhnika Kino i Televideniya (USSR); 1989 April; (4): 14-18.

Novakovskii, S.V. The open forum. Tekhnika Kino i Televideniya (USSR); 1988 February; (2): 3-6.

Novakovskiy, S.V. The outlook for the development of television systems with superhigh definition. Radiotekhnika (USSR); 1986 December; 41(12): 3-7.

Novakovskii, S.V. Some problems of creating the HDTV systems. Tekhinika Kino i Televideniya (USSR); 1983 June; (6): 53-5.

Nozawa, K.; Takada, T.; Shimazu, Y. High-speed time division switching using GaAs LSIs. Transactions of the Institute of Electronic Information Communication Engineers E (Japan); 1988 June; E71(6): 581-90.

Oda, E. et al. A 1920(H)*1035(V) pixel high-definition CCD image sensor. IEEE Journal of Solid State Circuits; 1989 June; 24(3): 711-17.

Ogura, Y. High-definition TV systems of TSUKUBA EXPO 85. Journal of the Institute of Television Engineers of Japan (Japan); 1985 July; 39(7): 597-8.

Ohara, S. Viewpoint of considering HA (home automation). Journal of the Institute of Electronic & Communication Engineers of Japan (Japan); 1985 September; 68(9): 947-50.

Ohgushi, K. et al. Sound system suitable for HDTV. NHK Laboratory

Note (Japan); 1988 February; (355): 1-11.

Ohgushi, K. et al. Subjective evaluation of multi-channel stereophony for HDTV. IEEE Transactions on Broadcasting; 1987 December; BC-33(4): 197-202.

Ohida, K. Project for improvement of television picture quality. Journal of the Institute of Television Engineering of Japan (Japan); 1986 May; 40(5): 348-9.

Ohno, S.; Sugiura, Y. Application of high definition television system for electronic imaging. Journal of Imaging Technology; 1986 October; 12(5): 261-6.

Ohtani, T.; Fujio, T.; Hamasaki, T. Subjective evaluation of picture quality for future high definition television. NHK Technical Journal (Japan); 1976; 28(4): 1-19.

Ohtsuka, Y.; Ninomiya, Y. A time-base corrector and other signal-processing equipment for a high-definition television video tape recorder. NHK Laboratory Notes (Japan); 1983 September; (290): 1-12.

Okada, K. A digital contour-corrector for a high-definition TV camera. NHK Laboratory Notes (Japan); 1985 January; (311):1-15.

Okada, K.; Mitsuhashi, T. HDTV-NTSC standards converters. NHK Laboratory Note (Japan); 1989 February; (366): 1-15.

Okano, F.; Kumada, J. HDTV hand-held camera using a 2/3-inch saticon. NHK Laboratory Note (Japan); 1986 November; (339): 1-9.

Oyamada, K.; Maeda, M.; Utsumi, Y. FDM optical fiber transmission for HDTV in broadcasting station. NHK Laboratory Note (Japan); 1987 April; (345): 1-10.

Ozaki, Y. Electron beam picture recording on 35mm film for HDTV. Image Technology (Great Britain); 1986 December; 68(12): 582-3.

Pacini, G.P.; Vitalone, RK. Direct TV reception by satellite in the present technological development. Elletronica e Telecomunicazioni (Italy); 1988; 37(1): 35-41.

Paskowski, M. Stereo TV takes center stage at NAB. Marketing & Media Decisions; 1984 June; 19(8): 76-8, 168, 170.

Pauchon, B. Synopsis of the operational and economic aspects of HDTV with a view to its standardization on a worldwide level. Revue Radiodiffusion et Television (France); 1988

October; 22(103): 6-9.

Pazarchi, M.; LoCiero, J.L. A matched-resolution wide aspect-ratio HDTV system. IEEE Transactions on Consumer Electronics; 1988 February; 34(1): 54-60.

Pech, E. Until high definition television arrives: topical developments in analogous receiver technology. Nachrichten Elektronik und Telematik (Germany); 1983 August-September; 37(8-9): 349-52.

Peitzmann, S. HDTV pilot project: challenge and expectancy. Funkschau (West Germany); 1988 October 21; (22): 34-8.

Peterson, T. Adding hustle to Europe's muscle. Business Week; 1989 June 16; (3110): 32, 34.

Peterson, T. Alain Gomez, France's high-tech warrior. Business Week; 1989 May 15; (3105 (Industrial/Technology Edition)): 100-106.

Phillimore, M. TV camera design. Middle East Electronics (Great Britain); 1986 January; 9(1): 24-7.

Phillips, G.J.; Harvey, R.V. High-definition television for satellite broadcasting. EBU Technical Review (Belgium); 1978 August; (170): 168-72.

Pirsch, P. Source coding of visual signals. IX. Methods with band separation. Nachrichtentechnische Zeitschrift NTZ (West Germany); 1985 February; 38(2): 99-100.

Plantholt, M.; Westerkamp, D.; Keesen, H.-W. Progressive scanning-the futuresafe solution for HDTV. Fernseh und Kino-Technik (West Germany); 1988 November; 42(11): 545-50.

Polonsky, J. Future high-fidelity television systems having higher definition. Rundfuntechnische Mitteilungen (Germany); 1981; 25(1): 12-15.

Polonsky, J. Future high definition television systems. Elektron - International (Austria); 1981; (3-4): 83-6.

Polonsky, J. Over a thousand lines: the next engineering goal? (Video signals). Television (Great Britain); 1980 January/February; 18(1): 13-17.

Polosin, L.A.; Roldugin, V.N.; Tatasova, T.A. Telecines for HDTV. Tekhnika Kino i Televideniya (USSR); 1989 January; (1): 18-22.

Powers, K.H. HDTV standards considerations for electronic

cinematography and post-production. SMPTE Journal; 1982 December; 91(12): 1153-7.

Powers, K.H. Techniques for increasing the picture quality of NTSC transmissions in direct satellite broadcasting. IEEE Journal of Selective Areas of Communication; 1985 January; SAC-3(1): 57-64.

Powers, K.H. A universal system for electronic film production. Fernseh und Kino-Technik (West Germany); 1985 February; 39(2): 81-6.

Pradenc, H. Stereophonic high resolution television. Micro Systems (France); 1988 September; (89): 167-72.

Prentiss, S. HDTV; high-definition television. Blue Ridge Summit, PA: Tab Books; 1989.

Prentiss, S. Time multiplexed compression: new dimensions in satellite TV. Satellite Communications; 1983 February; 7(2): 24-8.

Press, L. Thoughts and observations at the Microsoft CD-ROM conference. Communications of the ACM; 1989 July; 32(7): 784-88.

Price, J.N.; Silzars, A. eds. Selected papers from the 1985 SID International Symposium. Proceedings of the Society for Information Display; 1985; 26(4): 188 pages.

Price, J.N.; Silzars, A. eds. Selected papers from the 1985 SID International Symposium. Proceedings of the Society for Information display; 1986; 27(1): 188 pages.

Prokopova, Z. Telematics and new technology. Mechanizace Automatizace Administrativy (Czechoslovakia); 1988; (12): 476-8.

Przybyla, H.; Morita, T. Electron beam recording: the transfer of HDTV signals onto 35 mm film. Fernseh und Kino-Technik (West Germany); 1986 August; 40(8): 347-50.

Ramasastry, J. The road to HDTV. Satellite Communications; 1988 April; 12(4): 33-5.

Ramasastry, J.; Knights, G.F.; Cohen, H. Technical and regulatory aspects of satellite broadcasting: the US situation. Space Communications and Broadcasting (Netherlands); 1983 December; 1(4): 351-75.

Raths, D.; Jones, R.S. The DRAM game. InfoWorld; 1989 April 3; 11(14): 44-46.

Raven J.G. High definition MAC: the compatible route to HDTV.
IEEE Transactions on Consumer Electronics; 1988 February;
34(1): 61-3.

Reimers, U. Colour camera for HDTV systems. Fernseh und Kino-
Technik (West Germany); 1985 March; 39(3): 136-40.

Reimers, U. Origin and perceptibility of noise in a high
definition television (HDTV) camera. Frequenz (West
Germany); 1983 November/December; 37(11-12): 316-23.

Reimers, U.H. Resolution and noise-considerations for an HDTV
camera. SMPTE Journal; 1983 October; 92(10): 1036-40.

Reuber, C. Component signals and digital technique will improve
television. Funk-Technik (Germany); 1983 July; 38(7): 274-7.

Reuber, C. Future aims of TV technology. Funk-Technik (Germany);
1983 February; 38(2): 58-61.

Reuter, T. Motion adaptive downsampling of high definition
television signals. Proceedings of SPIE The International
Society for Optical Engineering; 1986; 594: 30-40.

Reuter, T. Standards conversion using motion compensation. Signal
Processing; 1989 January; 16(1): 73-82.

Reuter, T.; Hohne, H.D.; Ernst, M. Improved HDTV and TV
standards conversion. II. Fernseh und Kino-Technik (West
Germany); 1989; 43(3): 135-42.

Reuter, T.; Schamel, G.; Hilsky, R. Intensity linear processing
in a digital HDTV system. Fernseh und Kino-Technik (West
Germany); 1986 July; 40(7): 314-20.

Reynolds,D.; Keys, L. Signal distribution in tomorrow's
television plant. SMPTE Journal; 1986 October; 95(10):
1031- 3.

Rhodes, C.W. The B-MAC signal format as applied to 525 line TV
systems and to HDTV. IEEE Transactions on Consumer
Electronics; 1986 May; CE-32(2): 100-6.

Rhodes, C.W. Time division mulitplex of time compressed
chrominance for a compatible high definition television
system. IEEE Transactions on Consumer Electronics; 1982
November; CE-28(4): 592-603.

Rice, J.F. ed. HDTV; the politics, policies, and economics of
tomorrow's television. New York: Union Press; 1989.

Richter, H.-P. Interfaces and signal distribution in the
component studio. Fernseh und Kino-Technik (West

Germany); 1988 March; 42(3): 114-18.

Rieger, J.L. Raising the quality of NTSC. Digital Design; 1986 October 25; 16(12): 52-4.

Riemann, U.; Schonfelder, H.; Chmielewski, I. Methods of signal processing in a digital HDTV chromakey mixer. Fernseh und Kino-Technik (West Germany); 1988 June; 42(6): 259-64.

Riemann, U.; Wieben, W. Fast signal processing in a digital HDTV studio. Fernseh und Kino-Technik (West Germany); 1986 March; 40(3): 90-6.

Robert, P.; Lamnabhi, M.; Lhuillier, J.J. Advanced high-definition 50 to 60-Hz standards conversion. SMPTE Journal; 1989 June; 98(6): 420-24.

Robertson, B. Quantel launches Graphic Paintbox. Computer Graphics World; 1986 August; 9(8): 77-82.

Robertson, J. Congress unit hits AEA HDTV claim. Electronic News; 1989 August 7; 35(1170): 1-2.

Robinson, G.H. HDTV in the headend. CED; 1989 April; 15(4): 56, 58, 61.

Robson, T. High definition - the technical challenge. Electronics & Power; 1987 February; 33(2): 105-108.

Robson, T.S. Extended-definition television service. IEE Proceedings A (Great Britain); 1982 September; 129(7): 485- 92.

Roizen, J. Dubrovnik impasse puts high-definition TV on hold. IEEE Spectrum; 1986 September; 23(9): 32-7.

Roizen, J. High definition television demonstrations in US. Television (Great Britain); 1982 May/June; 19(3): 17-20.

Roizen-Telegen, J. High-definition television (HDTV)-standardization problems. Elletrotecnica (Italy); 1987 March; 74(3): 255-62.

Rosselevich, I. A. et al. Promising parameters of higher definition television system. Tekhnika Kino i Televideniya (USSR); 1987 January; (1): 5-11.

Ruprecht, J.; Lenth, J.; Johannes, K. Digital TV colour decoding. I. I/sup 2/C bus programming and luminance processing. Elektronik (Germany); 1987 March 20; 36(6): 121-4, 126-7.

Rzeszewski, T.S. A compatible high-definition television system. Bell Systems Technical Journal; 1983 September; 62(7, part 1): 2091-11.

Rzeszewski, T.S.; Pazarci, M.; LoCicero, J.L. Compatible high
definition television broadcast systems. IEEE Transactions
on Broadcasting; 1987 December; BC-33(4): 97-106.

Rzeszewski, T.S. Video coding in BISDN with a distribution rate
of approximately 135 Mb/s. International Journal of Digital
Analog Cabled Systems (Great Britain); 1988 January/March;
1(1): 33-40.

Sabatier, J.; Nasse, D. Standardization activities in HDTV
broadcasting. Signal Processing, Image Communication
(Netherlands); 1989 June; 1(1): 17-28.

Sabri, S.; Lemay, D.; Dubois, E. Modular digital video coding
architecture for present and advanced TV systems. SMPTE
Journal; 1989 July; 98(7): 504-11.

Sadashige, K. Video recording formats in transition. SMPTE
Journal; 1989 January; 98(1): 25-31.

Sakaue, K.; Hasegawa, S. Image processing and image
understanding. Journal of the Institute of Television
Engineers of Japan (Japan); 1988 July; 42(7): 684-7.

Salo, J.; Neuvo, Y.; Hameenaho, V. Improving TV picture quality
with linear-median type operations. IEEE Transactions on
Consumer Electronics; 1988 August; 34(3): 373-9.

Salvadorini, R. Broadcasting: the new services. Elletronica e
Telecomunicazioni (Italy); 1988; 37(1): 3-13.

Salvadorini, R. Direct satellite television and radiophony.
Elletronica e Telecommunicazioni (Italy); 1982
September/October; 31(5): 175-82.

Salvadorini, R. Direct satellite television and radio
transmission. Elletrotecnica e Telecominicazioni (Italy);
1987 December; 74(12): 1175-82.

Salvadorini, R. High definition television-chosen the standard.
Elletronica e Telecomunicazioni (Italy); 1988 November;
37(2): 50-4.

Salvadorini, R. TV standard for direct satellite broadcasting.
Elletronica e Telecomunicazioni (Italy); 1985 September-
October; 34(5): 191-8.

Salvadorini, R. Wideband networks-the head-end. Elletronica e
Telecomunicazioni (Italy); 1988 November; 37(2): 79-81.

Salvadorini, R.; D'Amato, P. High definition television.
Elletronica e Telecomunicazioni (Italy); 1986

September/October; 35(5): 187-96.

Samoilov, V.F.; Kononenko, I.A. Some peculiarities of resolution measurements for HDTV pickup tubes. Tekhnika Kino i Televideniya (USSR); 1987 June; (6): 30-4.

Sandbank, C.P.; Childs, I. The evolution towards high-definition television. Proceedings of IEEE; 1985 April; 73(4): 638-45.

Sandbank, C.P.; Moffat, M.E.B. High-definition television and compatibility with existing standards. SMPTE Journal; 1983 May; 92(5): 552-61.

Sansom, J.S. High definition television. Electronic Technology (Great Britain); 1986 March; 20(3): 268-71.

Sansom, J.S. A high definition television system. BKSTS Journal (Great Britain); 1986 May; 68(5): 255-7, 259-60.

Sata, N. Development of transmission systems. OEP Office Equipment & Products (Japan) supplement; 1987: 38-42.

Sato, S.; Kubo, T. Picture quality of high-definition television: 30-inch wide-screen color CRT display. NHK Laboratory Note (Japan); 1979 October; (241): 1-14.

Sauerburger, H. Wide and narrowband compatible single channel HDTV transmission. Fernseh und Kino-Technik (West Germany); 1987 January/February; 41(1-2): 23-32.

Sauerburger, H.; Stenger, L. Preprocessing and digital coding of HDTV-signals. Frequenz (West Germany); 1983 November/December; 37(11-12): 288-99.

Sauerburger, V.H.; Stenger, L. Transmissions of high-definition television signals using two satellite channels. Rundfunktechnische Mitteilungen (West Germany); 1984 September/October; 28(5): 235-40.

Savoskin, V.I.; Berezentseva, L.G. New zoom lenses for broadcast-quality TV cameras. Tekhnika Kino i Televideniya (USSR); 1987 December; (12): 55-61.

Sawada, K.; Murakami, H. HDTV (Hi-Vision). IV. Signal formats and transmission. Journal of the Institute of Television Engineers of Japan (Japan); 1988 September; 42(9): 951-8.

Schachlbauer, H. HDTV program storage in studios.. Rundfunktechnische Mitteilungen (West Germany); 1989 March/April; 33(2): 67-74.

Schachlbauer, H. HDTV studio recording. Rundfuntechnische Mitteilungen (West Germany); 1989 March/April;

33(2):67-74.

Schafer, R. High-definition television production standard-an opportunity for optimal color processing. SMPTE Journal; 1985 July; 94(7): 749-58.

Schafer, R.; Chen, S.-C. Considerations on new primaries for high definition television. IEEE Transactions on Consumer Electronics; 1988 August; 34(3): 513-22.

Schafer, R.; Golz, U. Investigations of subjective information transmission in a digital HDTV system. Fernseh und Kino-Technik (West Germany); 1985 February; 39(2): 73-80.

Schafer, R.; Kauff, P. HDTV colorimetry and gamma considering the visibility of noise and quantization errors. SMPTE Journal; 1987 September; 96(9): 822-33.

Schafer, R.; Kauff, P.; Golz, U. Subjective determination of methods of colour signal regeneration and processing with extended colour space and constant luminance principle. Fernseh und Kino-Technik (West Germany); 1987 May; 41(5): 201-6.

Schamel, G. Adaptive signal processing for digital HDTV 280 Mbit/s transmission. Signal Processing (Netherlands); 1988 October; 15(3): 335-50.

Schamel, G. Multi-dimensional prefiltering, sampling rate reduction and interpolation of HDTV signals. I. Frequenz (West Germany); 1988 October; 42(10): 284-8.

Schamel, G. Multidimensional prefiltering, sampling rate reduction and interpolation of HDTV signals. II. Frequenz (West Germany); 1988 November/December; 42(11-12): 300-4.

Schamel, G. Pre- and postfiltering of HDTV signals for sampling rate reduction and display up-conversion. IEEE Transactions on Circuits Systems; 1987 November; CAS-34(11): 1432-9.

Schamel, G.; Hahn, M. Noise reduction in high-definition TV cameras. Fernseh und Kino-Technik (West Germany); 1988 July; 42(7): 312-18.

Schiffler, W. HDTV-productions in the EUREKA-standard. Fernseh und Kino-Technik (West Germany); 1989 43; 6(295-8).

Schmidt, M. The picture tube has a future. Funkschau (West Germany); 1988 October 21; (22): 56-8 supplement.

Schneider, A. A system generating high-resolution animation to

HDTV film. SMPTE Journal; 1986 August; 95(8): 796-8.

Schneider, W.C.; Resor, G.L. High-volume production of large full-color liquid crystal displays. Information Display; 1989 February; 5(2): 6p.

Schonfelder, H. Digital control technology in the HDTV television studio. Fernseh und Kino-Technik (West Germany); 1987 June; 41(6): 237-42.

Schonfelder, H. Digital HDTV video mixing. Fernseh und Kino-Technik (West Germany); 1989; 43(6): 301-4, 306-7.

Schonfelder, H. From the component studio to high definition television. Nachrictentechnische Zeitschrift (West Germany); 1988 July; 41(7): 406-9.

Schonfelder, H. Possibilities of quality improvement in today's television systems. Fernseh und Kino-Technik (West Germany); 1983 May; 37(5): 187-95.

Schonfelder, H. Processing problems in the HDTV-studio. Fernseh und Kino-Technik (West Germany); 1985 March; 39(3): 109-14.

Schreiber, W.F. Advanced television systems for the United States: getting there from here. SMPTE Journal; 1988 October; 97(10): 847-51.

Schreiber, W.F. et al. Channel-compatible 6-MHz HDTV distribution systems. SMPTE Journal; 1989 January; 98(1): 5-13.

Schreiber, W.F. HDTV technology: advanced television systems and public policy options. Telecommunications; 1987 November; 21(11 (North American Edition)): 37-42.

Schreiber, W.F. Improved television systems: NTSC and beyond. SMPTE Journal; 1987 August; 96(8): 734-44.

Schreiber, W.F.; Lippman, A.B. Reliable EDTV/HDTV transmission in low-quality analog channels. SMPTE Journal; 1989 July; 98(7): 496-503.

Schroder, H. On the definition and measurement of the resolution of TV picture reproduction equipment. Fernseh und Kino-Technik (West Germany); 1985 April; 39(4): 163-8.

Schroder, H.; Huerkamp, G. System and circuit structures for HDTV picture presentation with 100 Hz picture change frequency. ITG-Fachberichte (West Germany); 1988 103; (159-65).

Schwarz, H. Videopaths, distribution and interfaces in the digital studio. Fernseh und Kino-Technik (West Germany);

1988 August; 42(8): 351-6.

Segawa, K. et al. 20-inch hi-vision TV monitor. National Technical Reports (Japan); 1988 October; 34(5): 62-65.

Seghers, F. Television makers are dreaming of a wide crispness. Business Week; 1987 December 21; (3031 (Industrial/Technology Edition)): 108-09.

Sell, G. Satellites ready and waiting for HDTV. CED; 1989 March; 15(3): 50-6.

Sell, G. Tomorrow's converters: tuning in the future. CED; 1989 June; 15(7): 86-93.

Shelswell, P.; Dosch, C. Satellite broadcasting of HDTV: some basic considerations. EBU Technology Review (Belgium); 1986 October; (219): 315-25.

Shibata, Y. Recent development on broadcasting technology-high definition television II. AEU (Japan); 1984 December: 110- 16.

Shibata, Y. et al. Standard cell for HDTV. Sharp Technical Journal (Japan); 1988 October; (40): 56-60.

Shibaya, H. VTR for high-definition television. Journal of the Institute of Television Engineers of Japan (Japan); 1985 April; 39(4): 316-20.

Shibaya, H. et al. Recording and reproducing equipment for the high-definition television. Journal of the Institute of Television Engineers of Japan (Japan); 1985 August; 39(8): 675-9.

Shiga, F. Mobility, manipulation and new media mark developments in Japan's broadcast industry. JEE (Japan); 1983 April; 20(196): 28-31.

Shimada, J.; Kato, M. Trend in optical disc memory. Journal of the Society of Instrumentation & Control Engineers (Japan); 1985 June; 24(6): 533-8.

Shimazu, Y.; Nozawa, K.; Takada, T. High-speed time division switching system using GaAs devices. Electronic Communications Laboratory Technical Journal (Japan); 1988; 37(12): 821-30.

Shimazu, Y.; Takada, T. High-speed time switch using GaAs LSI technology. IEEE Journal of Select Areas of Communication; 1986 January; SAC-4(1): 32-8.

Shimizu, T. et al. 10-bit 20-MHz two-step parallel A/D converter with internal S/H. IEEE Journal of Solid-State Circuits; 1989 February; 24(1): 13-20.

Shimizu, Y. Students evaluate TV lecture system. Business Japan (Japan); 1986 March; 31(3): 81, 83.

Shuni, C.; Chen, T.C. A flexible format video sequence processing simulation system. Proceedings of SPIE The International Society for Optical Engineering; 1986; 707: 116-203.

Silverberg, M. HQTV systems-a comparison. Fernseh und Kino-Technik (West Germany); 1988 October; 42(10): 483-4, 497-8.

Simonsen, S.O. Compatibility of NTSC decoders with HDTV signals. IEEE Transactions on Consumer Electronics; 1986 February; CE-32(1): 62-7.

Slamin, B. Plain man's guide to Eureka 95: compatible high definition television. Television (Great Britain); 1987 May/June; 24(3): 120-1.

Smith, C.W.; Dumbreck, A.A. 3D TV: the practical requirements. Television (Great Britain); 1988 January/February; 25(1): 9- 15.

Smith, L. Can consortiums defeat Japan? Fortune; 1989 June 5; 119(12): 245-54.

Sochor, J. Problems of the magnetic tape recording of broadband signals. Fernseh und Kino-Technik (West Germany); 1983 May; 37(5): 197-202.

Solomon, D. HDTV: television's next generation. Marketing & Media Decisions; 1989 March; 24(3): 109-111.

Sommerfeldt, H. Colour picture tubes. Radio Fernsehen Elektronik (East Germany); 1988; 37(12): 758-60.

Soroka, E.Z. Multiline television (HDTV). Tekhnika Kino i Televideniya (USSR); 1983 May; (5): 42-51.

Stamberger, A. Alternative introduction strategies for improving TV picture quality. Elektroniker (Switzerland); 1987 September; (9): 137-44.

Stammnitz, P.; Hofker, U. Digital HDTV system. Fernseh und Kino-Technik (West Germany); 1989; 43(5): 227-33.

Stanton, J.A.; Stanton, M.J. Bibliography: video production technologies. SMPTE Journal; 1986 August; 96(8): 762-69.

Stanton, J.A.; Stanton, M.J. Video recording: a history. SMPTE Journal; 1987 March; 96(3): 253-63.

Stenger, L. Digital coding of TV signals for ISDN-B applications.
    IEEE Journal on Select Areas of Communication; 1986 July;
    SAC-4(4): 514-28.
Stenger, L. HDTV-a new high definition television system.
    Computer Graphics Forum (Netherlands); 1985 June;
    4(2): 117- 25.
Stenger, L. HDTV: the European alternative. Funkschau (West
    Germany); 1988 July 29; (16): 26-8.
Stephens, G.M. Technology: being pulled by markets. Satellite
    Communications; 1988 May; 12(5): 28-9.
Stevens, M. Audio visual: projection's progress. Marketing
    (Great Britain); 1987 April 30; 29(5): 43-6.
Stoddard, R. 'Dramatic changes' in satellite broadcasting.
    Satellite Communications; 1988 November; 12(11): 28-30.
Stoddard, R. The road back in HDTV. Satellite Communications;
    1989 April; 13(4): 14-18.
Stoddard, R.; Grimes, A.J. U.S. DBS on the comeback trail.
    Satellite Communications; 1988 July; 12(7): 15-19.
Stollenwerk, F. Compatible improved TV systems and their
    subjective evaluation. Fernseh und Kino-Technik (West
    Germany); 1985 February; 39(2): 64-72.
Stollenwerk, F. Viewing distance of future TV systems. Fernseh
    und Kino-Technik (West Germany); 1986 May; 40(5)188-92.
Storey, R. HDTV motion adaptive bandwidth reduction using DATV.
    Television (Great Britain); 1987 March/April; 24(2):
    84-8, 90-1.
Strain, R.A. The shape of screens to come (TV). SMPTE Journal;
    1988 July; 97(7): 560-7.
Stumpf, R.J. A film studio looks at HDTV. SMPTE Journal; 1987
    March; 96(3): 247-52.
Stumpf, R.J. A film studio looks at HDTV. Image Technology (Great
    Britain); 1986 September; 68(9): 428-29, 431-33.
Sugimori, Y.; Nishizawa, T. EDTV. Journal of the Institute of
    Television Engineers of Japan (Japan); 1988 July; 42(7):
    659-60.
Sugimoto, A.; Tsuji, M.; Shikina, C. The digital video effects
    for HDTV. NEC Technical Journal (Japan); 1985 May;
    38(5): 33-6.
Sugimoto, A.; Tsuji, M.; Shikina, C. The NTSC to HDTV up-

converter. NEC Technical Journal (Japan); 1985 May; 38(5): 37-40.

Sugimoto, M.; Mitsuhashi, T.; Tsuzikawa, K. High-definition television. Journal of the Institute of Television Engineers of Japan (Japan); 1986 July; 40(7): 614-18.

Sugimoto, M.; Mitsuhasi, T.; Yoshino, T. New technologies affecting radio and TV broadcasting. Journal of the Institute of Electronic and Communications Engineers (Japan); 1984 July; 67(7): 757-62.

Sugimoto, M.; Watanabe, S. The Cosmic Hall of the TSUKUBA EXPOCenter (HDTV system). Journal of the Institute of Television Engineers of Japan (Japan); 1985 July; 39(7): 599-605.

Sugimoto, T. et al. A high-quality 40-inch CRT color display with high resolution and definition. Mitsubishi Denki Giho (Japan); 1985; 59(3): 221-3.

Sugiura, T. Giant screen movies and systems of TSUKUBA EXPO 85. Journal of the Institute of Television Engineers of Japan (Japan); 1985 July; 39(7): 628-35.

Sugiura, Y. Laser-beam film recording for high-definition television. NHK Laboratory Note (Japan); 1988 August; (360): 1-11.

Sugiura, Y.; Ishida, T.; Ozaka, Y. Film systems for HDTV. Journal of the Institute of Television Engineers of Japan

Sunada, K. et al. High picture quality TV receiver with IDTV system. IEEE Transactions on Consumer Electronics; 1988 November; 34(4): 856-65.

Sutherland, D. Presentation imaging for the 1990's. Business Marketing; 1989 June; 74(6): 62-72.

Suzuki, C. Displays for new media systems. Journal of the Institute of Television Engineers of Japan (Japan); 1985 January; 39(1): 96-7.

Suzuki, J. Future prospects for the high definition video system. BKSTS Journal (Great Britain); 1983 October; 65(10): 556-9.

Suzuki, N. et al. Multiplexing of synthetic motion signals in a color encoder for scan conversion. Journal of the Institute of Television Engineers of Japan (Japan); 1988 September; 42(9): 966-72.

Suzuki, N. et al. NTSC scan conversion using motion adaptive processings. NEC Research & Development (Japan); 1985 April; (77): 38-44.

Suzuki, Y. et al. 40 inch CRT display for Hi-Vision. Journal of the Institute of Television Engineers of Japan (Japan); 1985 December; 39(12): 1168-75.

Suzumori, S. et al. Barium ferrite tape application to high-definition VTRs. Toshiba Review (International Edition) (Japan); 1985 Winter; (154): 23-6.

Tadokoro, Y. What Japan is doing with DBS. Cable & Satellite Europe (Great Britain); 1985 January; (1): 10-22.

Taguchi, S. High-definition TV receiver-a newly-developed MUSE receiver and 40-inch CRT display. Toshiba Review International Edition (Japan); 1985 Summer; (152): 38-41.

Taguchi, S..; Kataoka, T. Display for high-definition TV. Journal of the Institute of Television Engineers of Japan (Japan); 1985 August; 39(8): 685-9.

Takahashi, N. Digital still HDTV disc system. IEEE Transactions on Consumer Electronics; 1988 February; 34(1): 64-71.

Takegahara, T. et al. Sound transmission system using baseband time division multiplexing for MUSE system HDTV. NHK Laboratory Note (Japan); 1988 April; (358): 1-12.

Takegahara, T.; Tanabe, H.; Suganami, H. Sound transmission for HDTV using baseband multiplexing into MUSE video signal. IEEE Transactions on Broadcasting; 1987 December; BC-33 (4): 188-96.

Takehara, N.; Kotera, H. Color reproduction methods for hard-copy printer. Electrophotography (Japan); 1985; 24(1): 60-7.

Talmi, M.; Jara, E. Parallel arithmetic units for HDTV component systems. ITG-Fachberichte (West Germany); 1988; 103: 167-71.

Tanabe, H. High quality sound program coding-broadcast quality sound program coding. Journal of the Institute of Television Engineers of Japan (Japan); 1986 August; 40(8): 706-8.

Tanaka, M. Introduction of audio-video products. Journal of the Institute of Television Engineers of Japan (Japan); 1983 June; 37(6): 503-4.

Tanaka, T. Practical applications of SHF TV broadcasting. JEE (Japan); 1980 October; 17(166): 39-42.

Tanaka, Y. et al. HDTV-PAL standards converter. NHK Laboratory

Note (Japan); 1986 January; (326): 1-15.

Tanaka, Y. et al. HDTV-PAL standards converter using motion compensation technique and its converted picture quality. Transactions of the Institute of Electronic Information Communication Engineers. D (Japan); 1987 August; J70D(8): 1535-45.

Tanaka, Y. et al. Hi-vision-PAL standards converter (HDTV). NHK Technical Journal (Japan); 1987; 39(2): 112-29.

Tanimoto, M.; Chiba, N.; Kageyama, M. An experimental system of one-dimensional TAT bandwidth compression for television signals. Electronic Communication of Japan. 1. Communications (USA); 1989 February; 72(2): 77-86.

Tanimoto, M. et al. TAT (time-axis transform) bandwidth compression system of picture signals. IEEE Transactions on Communications; 1988 March; 36(3): 347-54.

Tanimoto, M. et al. TAT (time-axis transform) bandwidth compression system for high-definition television. Journal of the Institute of Television Engineers of Japan (Japan); 1985 October; 39(10): 934-40.

Tanimoto, M.; Mori, T. A hybrid scheme of subsampled DPCM and interpolative DPCM for the HDTV coding. Transactions of the Institute of Electronic Information and Communication Engineers (Japan); 1987 July; E70(7): 611-13.

Tanimoto, M.; Yamada, A. Image restoration for complementary subsampling system. Transactions of the Institute of Electronic Information Communication Engineers. B (Japan); 1988 December; J71B(12): 1511-16.

Tazaki, S.; Yamada, Y. Systematization on image coding and its application. Journal of the Institute of Electronic Information Communication Engineers (Japan); 1988 July; 71(7): 663-8.

Tchen, H.; Paret, D. Which display devices for high-definition television? Onde Electronique (France); 1989 July/August; 69(4): 35-42.

Tchen, H.; Paret, D. Which display systems for high-definition television? Revue Radiodiffusion et Television (France); 1988 October; 22(103): 10-17.

Tejerina Garcia, J.L. High definition television. Mundo Electronico (Spain); 1986; (162): 113-20.

Tepe, R. Analysis of fluid light valve control layers for high-
    definition television large picture projection. Journal of
    Applied Physics; 1985 April 1; 57(7): 2355–60.
Terashima, T. Switching transistor for TV horizontal deflection.
    Sanken Technical Reports (Japan); 1985 November;
    17(1): 9-26.
Tetzner, K. Entertainment electronics in Japan: firmly in saddle
    at home. Funkschau (West Germany); 1987 May 8; (10):
    55, 58.
Tetzner, K. Picture tubes: projection comes of age. Funkschau
    (West Germany); 1989 May 5; (10): 67-9.
Tetzner, K. The ways towards better television pictures. When
    will the new standards come? Elektronik (West Germany);
    1987 August 21; 36(17): 76-80.
Tetzner, K.; Radke, G.L. Satellite television: D2-Mac and the
    consequences. Europe at the threshold of a new TV era.
    Funkschau (West Germany); 1985 August 30; (18): 59-63.
Thomas, G.A. HDTV bandwidth reduction by adaptive subsampling
    and motion-compensation DATV techniques. SMPTE
    Journal; 1987 May; 96(5, part 1): 460-5.
Thorpe, L.J. Sony technology yields practical system for super
    slow-motion video reproduction. JEE (Japan); 1985 March;
    22(219): 34-8, 96.
Thorpe, L.J.; Ozaki, Y. HDTV electron beam recording. SMPTE
    Journal; 1988 October; 97(10): 833-43.
Thryft, A. Domestic DRAM makers applaud U.S. memories.
    Computer Design; 1989 July 17; 28(14): 1-2.
Titus, J. How to succeed in the 1990's; global perspective,
    innovation, and manufacturing where you sell will be keys.
    EDN; 1989 June 29; 34(12A): S76-80.
Togawa, A.; Takeuchi, M.; Sawakuri, T. Overall operation results
    for the INS model system. Review of the Electrical
    Communication Laboratories (Japan); 1986 March; 34(2):
    191- 97.
Tohara, H. Trends of HDTV. NEC Technical Journal (Japan); 1985
    May; 38(5): 25-7.
Tomlinson, M.; Ward, C.; Rendle, M. Digital pseudo-analogue
    satellite TV transmission system. IEE Proceedings F (Great
    Britain); 1986 July; 133(4): 384-98.

Tomoda, K. Introduction of giant-screen display by CRT projection system. NEC Technical Journal (Japan); 1985 August; 38(8): 2-4.

Tomoda, K. A view of new television age. NEC Technical Journal (Japan); 1988 November; 41(13): 3-6.

Tomoda, K. et al. 400-inch video projector for HDTV display. NEC Technical Journal (Japan); 1985 May; 38(5): 47-54.

Tomoda, K. et al. 400-inch video projector and control system. NEC Technical Journal (Japan); 1985 August; 38(8): 12-18.

Tomoda, K. et al. Automatic convergence alignment system for video projectors in parallel operation. NEC Technical Journal (Japan); 1985 August; 38(8): 19-23.

Tonge, G. Signal processing for higher-definition television. IBA Technical Review (Great Britain); 1983 November; (21): 13- 26.

Tonge, G.J. The compatible delivery of HDTV to the home. Television (Great Britain); 1988 November/December; 25(6): 288-92.

Tonge, G.J. Image processing for higher definition television. IEEE Transactions on Circuits Systems; 1987 November; CAS- 34(11): 1385-98.

Toth, A.G.; Tsinberg, M.; Rhodes, C.W. Hierarchical NTSC compatible HDTV system architecture-a North American perspective. International Journal of Digital Analog Cabled Systems (Great Britain); 1988 April/June; 1(2): 65-72.

Toth, A.G.; Tsinberg, M.; Rhodes, C.W. NTSC compatible high definition television emission system. IEEE Transactions on Consumer Electronics; 1988 February; 34 140-7.

Toth, A.G.; Tsinberg, M. Hierarchical evolution of high definition television. IEEE Transactions on Broadcasting; 1987 December; BC-33(4): 124-9.

Toyama, T. et al. Optical videodisc for high-definition television by the MUSE. SMPTE Journal; 1986 January; 95 (1 part 1): 25-9.

Toyama, T. et al. Optical video disc for Hi-vision. Sanyo Technical Review (Japan); 1987; 19(1): 24-33.

Troller, G.; Ehrke, H.-J.; Sinnig, T. Are CRT projectors HDTV compatible? Fernseh & Kino-Technik (West Germany); 1985 September; 39(9): 421-4.

Truskalo, W. Resonant degaussing for TV and high definition color monitors. IEEE Transactions on Consumer Electronics; 1986 November; CE-32(4): 713-22.

Trzebunia-Siwicka, W. High definition television and problems connected with transmission. Przeglad Telekomunikacyjny (Poland); 1986; 59(6): 166-72.

Tsinberg, M. ENTSC two-channel compatible HDTV system. IEEE Transactions on Consumer Electronics; 1987 August; CE-33 (3): 146-53.

Tsuboi, T.; Asatani, K.; Miki, T. Fiber-optic HDTV transmission technology. SMPTE Journal; 1985 January; 94(1): 4-10.

Tsuboi, T. et al. Fiber optic high-definition television transmission system. Electronic Communications Laboratory Technical Journal (Japan); 1984; 33(11): 2693-704.

Tsuboi, T.; Obara, H.; Asatani, K. Design and characteristics of high-definition television transmission system. Transactions of the Institute of Electronic & Communication Engineers of Japan Part B (Japan); 1985 December; J68B(12): 1405-12.

Tsuboi, T.; Obara, H.; Asatani, K. Fiber optic high-definition television on transmission system. Review of the Electronic Communication Laboratory (Japan); 1985 July; 33(4): 609 14.

Tsuda, T.; Hiraoka, M. Efficient video bandwidth compression. Fujitsu Scientific & Technical Journal (Japan); 1986 Autumn; 22(4): 355-66.

Tsujihara, S. et al. 50-inch rear-projection display for Hi-Vision. National Technical Reports (Japan); 1987 April; 33(2): 224-32.

Tsukamoto, J.; Hayashida, H. 12 GHz experimental station-high-definition TV experiment in TSUKUBA. Journal of the Institute of Television Engineers of Japan (Japan); 1985 July; 39(7): 605-8.

Tunmann, E. Innovative design of a commercial insertion system. CED; 1988 September; 14(9): 58-69.

Uba, T. et al. 16:9 aspect ratio 38V-high resolution trinitron CRT for HDTV. IEEE Transactions on Consumer Electronics; 1988 February; 34(1): 85-90.

Udagawa, H. NHK: a tale of dreams. Tokyo Business Today (Japan); 1988 November: 46-48.

Uehara, T.; Sakai, T.; Mochizuki, T. A wideband and low noise reproduction system for VCR. NHK Laboratory Note (Japan); 1987 January; (342): 1-11.

Uematsu, K. Japan designs a reprocessing pilot plant for remote maintenance. Nuclear Engineering International; 1987 February; 32(391): 40-41, 44.

Ueno, K. et al. High definition TV decoder. Sharp Technical Journal (Japan); 1986; (34): 71-6.

Umemoto, M. et al. An experimental 648 Mbit/s HDTV digital VTR. IEEE Transactions on Broadcasting; 1987 December; BC-33 (4): 210-13.

Umemoto, M. et al. High data rate recording for HDTV digital VTR. IEEE Transactions on Magnetics; 1988 November; 24(6): 2407- 9.

Umcmoto, M. et al. Recording channel electronics for HDTV digital VTR. IEEE Transactions on Magnetics; 1987 September; MAG- 23(5 part 2): 3173-5.

Umemoto, M.; Mita, S.; Eto, Y. Experiments of digital recording for high definition television signals. Journal of the Institute of Television Engineers of Japan (Japan); 1986; 40(11): 1120-25.

Unnai, T.; Goto, N.; Takashaki, Y. Newly developed high-sensitivity HDTV camera tube. Hitachi Review (Japan); 1989 April; 38(2): 121-6.

Uno, J. Recent trends of television broadcasting engineering. Journal of the Institute of Television Engineers of Japan (Japan); 1987 January; 41(1): 3-6.

Urata, Y.; Honjo, S. Color motion picture film for laser recording. Scientific Publications of the Fuji Photo Film Company, Ltd. (Japan); 1988; (33): 9-14.

van der Klugt, C.J. New television standards, revolution or evolution? Image Technology (Great Britain); 1986 September; 434-5, 437-9, 460.

van Houten, S. Applied research-the source of innovation in consumer electronics. Philips Technical Review (Netherlands); 1988 December; 44(6): 180-9.

Van Lewen, K. Will consumers buy HDTV? Satellite Communications; 1989 January; 13(1): 23-24.

Van Loan, J. EIA Multiport: an idea whose time has come. CED; 1989 June; 15(7): 120-3.

Van Tilburg, J. TV at the beginning of a new evolution. Funkschau (West Germany); 1988 October 21; (22): 67-8 supplement.

Vaughan, T.; McClure, II. Reflection and ghosts in a multitower environment. IEEE Transactions on Broadcasting; 1989 March; 35(1): 8-22.

Veillard, J. The HD-MAC system: radio characteristics. Revue Radiodiffusion et Television (France); 1988 July/September; 22(102): 24-7.

Veillard, J. The HDMAC system: radio characteristics. Onde Electronique (France); 1989 July/August; 69(4): 20-5.

Visintin, F. RAI electronic film project. EBU Technical Review (Belgium); 1988 April; (228): 52-59.

Vogt, C. Two dimensional digital aperture correction for HDTV cameras. 1. Processing gradation precorrected camera signals. Fernseh und Kino-Technik (West Germany); 1989; 43(7): 356-62.

von Reden, W.; Sinnig, T. Simulation of the modulation transfer function of a projecting optical system. Applied Optics; 1987 October; 26(20): 4416-22.

Voore, T. The incredible Scophony receiver. Electronic & Wireless World (Great Britain); 1987 May; 93(1615): 515-17.

Wakui, K. Broadcasting new-media and HA. Journal of the Society of Instrumation & Control Engineering; 1984 November; 23(11): 927-31.

Walf, G. et al. Integrated services broadband transmission system with high-rate optical subscriber lines. Nachrichtentechnische Zeitschrift (West Germany); 1986 May; 39(5): 302-4, 306, 308, 310-11.

Walker, S.L. The contractor that's coming in from the cold: science applications moves beyond its secret Pentagon work. Business Week; 1988 November 14; (3079 (Industrial/Technology Edition)): 150, 154.

Wassiczek, N. Trends and developments in television production. Elektrotechnik und Maschinenbau (EUM) (Austria); 1982 November; 99(11): 482-3.

Watanabe, T. et al. A high-definition television receiver. Mitsubishi Denki Giho (Japan); 1986; 60(11): 9-13.

Weber, D.M. Digital circuits point towards better TV sets. Electronics Weekly; 1984 August 13; 57(18): 49-53.

Webers, J. 24 or 30 picture/s picture frequency? Is a new standard for movie films significant? Fernseh und Kino-Technik (West Germany); 1988 October; 42(10): 451-8.

Webers, J. Film and video techniques: new directions for the production of programmes for the 1990s. Fernseh und Kino-Technik (West Germany); 1986 August; 40(8): 351-2, 354-6.

Weckenbrock, H.; Wedam, W. ACTV: advanced compatible television. Fernseh und Kino-Technik (West Germany); 1988 July; 42(7): 305-11.

Wedam, W.F. Future trends in television. IEEE Transactions on Consumer Electronics; 1988 May; 34(2): 279-84.

Weimer, P.K. Television image sensors - tubes versus solid-state. Journal of Imaging Technology; 1986 October; 12(5): 244-57.

Wendland, B. Picture scanning and subjective picture quality. Fernseh und Kino-Technik (West Germany); 1985 February; 39(2): 56-63.

Wendland, B. Strategies for high definition television systems. Fernseh und Kino-Technik (West Germany); 1981 September; 35(9): 325-32.

Wendler, K.-P. Concepts for picture raster conversion in EDTV and HDTV. Fernseh und Kino-Technik (West Germany); 1988 August; 42(8).

Wendtlandt, B. et al. Television for the next millennium. Radio Mentor Electronik (Germany); 1980; 46(1-2): 26-7.

Westerink, J.H.D.M.; Roufs, J.A.J. Subjective image quality as a function of viewing distance, resolution, and picture size. SMPTE Journal; 1989 February; 98(2): 113-19.

Whitten P. Japanese new media craze (DBS project). Cable and Satellite Europe (Great Britain); 1984 December; (12): 9-12.

Wiegner, K.K. Last chance? Forbes; 1988 May 30; 141(12): 58-64.

Willcock, C.A. Service implications of fiber networking. IEEE Transactions on Consumer Electronics; 1989 May; 35(2): 92-6.

Wilson, A.C. Imaging and graphics: solid state sensors capture bigger images. Digital Design; 1986 December; 16(14): 43-45.

Windram, M.D.; Tonge, G.J.; Hills, R.C. D-MAC/packet transmission system for satellite broadcasting in the United Kingdom. EBU Technical Review (Belgium); 1988 February; (227): 21-28.

Windram, M.D.; Morcom, R.; Hurley, T. Extended-definition MAC. IBA Technical Review (Great Britain); 1983 November; (21): 27-41.

Windram, M.D.; Tonge, G.J.; Hills, R.C. Satellite broadcasting in the United Kingdom. IBA Technical Review (Great Britain); 1988 November; (24): 5-10.

Wise, D. Thomson's French revolution. Business (Great Britain); 1989 June: 48-56.

Wu, D. Technical basis of planning the fixed-satellite service. A summary of the CCIR intersessional work. Telecommunication Journal; 1988 August; 55(8): 539-47.

Wurl, W. A new professional video recorder of the GPR family for high definition TV recording with increased bandwidth. Grundig Technische Informationer (West Germany); 1978; 25(5): 293-9.

Yagi, N. et al. An architecture of real-time video signal processing LSI-Picot. Transactions of the Institute for Electronic Communications Engineers. C-II (Japan); 1989 May; J72C-II(5): 346-53.

Yamada, M.; Fukuda, T. Analysis of television pictures using eye movement. Journal of the Institute of Television Engineers of Japan (Japan); 1986 February; 40(2): 121-8.

Yamada, M.; Fukuda, T. Development of a gazing point analyser (vision analyser) and its application to broadcasting. NHK Laboratory Note (Japan); 1985 July; (317): 1-15.

Yamada, M.; Fukuda, T. Quantitive evaluation of eye movements as judged by sight-line displacements: comparison of eye-movement patterns during 525-line and HDTV viewing, and evaluation of moving images. SMPTE Journal; 1986 December; 95(12): 1230-41.

Yamaguchi, A. Color TV: larger screens no longer mean poor pictures. JEE Journal of Electronic Engineering (Japan); 1987 September; 24(249): 28-31.

Yamaji, M. Display technology trends. Mitsubishi Denki Giho (Japan); 1989; 63(3): 2-4.

Yamanaka, N. et al. 1 Gbit/s, 32 multiplied by 32 high-speed space-division switching module for broadband ISDN using SST LSIs. Electronics Letters (Great Britain); 1989 June 22; 25(13): 831-33.

Yamanaka, N.; Miyanaga, H.; Yamamoto, Y. High-speed time division switch for 32-Mbit/s bearer rate signals. IEEE Journal of Select Areas of Communication; 1987 October; SAC-5(8): 1249- 55.

Yamanaka, N.; Miyanaga, H.; Yamamoto, Y. Newly structured 512 Mbit/s high-speed time-division switch. Electronic Letters (Great Britain); 1986 October 9; 22(21): 1094-6.

Yamasaki, R. Portable color TV with a built-in stereo micro radio cassette recorder. Toshiba Review (International Edition) (Japan); 1983 Summer; (144): 29-31.

Yamazaki, E. Applications and large picture screens of CRT displays expanding. JEE (Japan); 1982 September; 19(189): 88-9.

Yamazaki, E.; Ueda, T.; Otsuka, S. Experimental display and pick-up devices for a high-definition TV system. Proceedings of S.I.D.; 1982; 23(3): 129-34.

Yamazaki, J. et al. New apparatus for measuring photoconductive characteristics linked to vacuum evaporation equipment. NHK Laboratory Note (Japan); 1986 October; (337): 1-11.

Yamazaki, K.; Fukugawara, T.; Soejima, T. Fiber-optic subscriber loop network. Fujitsu (Japan); 1986; 37(6): 473-80.

Yanase, S. et al. Hi-vision MUSE/NTSC converter. Sanyo Technical Review (Japan); 1989 June; 21(2): 40-8.

Yashima, Y.; Sawada, K. Adaptive intraframe/interframe coding for HDTV signals by using extrapolative and interpolative prediction. Transactions of the Institute of Electronic Information & Communication Engineers. B (Japan); 1987 January; J70B(1): 96-104.

Yasuda, H.; Kishino, F. Future prospects of visual communications network. Journal of the Institute of Television Engineers of Japan (Japan); 1988 June; 42(6): 538-45.

Yasumoto, Y. et al. An extended definition television system using quadrature modulation of the video carrier with inverse Nyquist filter. IEEE Transactions on Consumer Electronics; 1987 August; CE-33(3): 173-80.

Yasumoto, Y. et al. An NTSC compatible wide screen television system for terrestrial broadcasting. National Technical Reports (Japan); 1988 October; 34(5): 38-46.

Yip, W.C.; Kongable, L. Advanced digital television system. IEEE

Transactions on Consumer Electronics; 1986 November; CE-32(4): 743-53.

Yokoo, T. The DBS system in Japan. AEU (Japan); 1986 October; (129): 55-9.

Yokoo, T. Reception system for direct broadcasting satellite and high definition TV in Japan. AEU (Japan); 1986 January: 24- 30.

Yokoyama, K. et al. Development of a VTR for the high-definition television. NHK Technical Journal (Japan); 1986; 38(2): 1-54.

Yokozawa, M. Hi-Vision broadcasting (HDTV). JEE Journal of Electronic Engineering (Japan) supplement; 1987: 63-4.

Yokozawa, M. Hi-vision broadcasting. OEP Office Equipment & Products (Japan) supplement; 1987: 63-4.

Yoshimoto, S. et al. A study on 22 GHz-band multibeam satellite broadcasting system. Transactions of the Institute of Electronic and Communications Engineers of Japan. Part B (Japan); 1986 November; J69B(11): 1258-66.

Yoshizawa, A. The current status of high-definition television program productions. Journal of the Institute of Television Engineers of Japan (Japan); 1985 August; 39(8): 700-4.

Yoshizawa, H.; Fushiki, K. Digital technology applied to high-resolution television for home electronic devices of the next generation. Nikkei Electronics (Japan); 1986; (403): 123-43.

Yoshizawa, H.; Matsunaga, N. IDTV is on the market. Nikkei Electronics (Japan); 1988; (451): 169-75.

Yoshizawa, T. Standardization of EDTV system aiming to start broadcasting in April, 1989. Nikkei Electronics (Japan); 1987; (436): 149-61.

Yoshizawa, Y. International standard on HDTV. Nikkei Electronics (Japan); 1987; (427): 97-112.

Young, J.; Cohen, C.L.; Iverson, W.R. At last, the TV picture gets sharper. Electronics; 1987 October 15; 60(21): 113-114.

Young, R.W. EUREKA and production standards for HDTV. Image Technology (Great Britain); 1987 December; 69(12): 527-9.

Yuyama, I.; Yano, S. Picture-quality of high definition television with gamma correction at the receiver. Transactions of the Institute of Electronic and Communications Engineers of Japan Part A. (Japan); 1986 March; J69A(3): 391-9.

Ziemer, A. Broadcasting satellites and the ISDN: new networks
for television of tomorrow. Funkschau (West Germany); 1986
October 24; (22): 32-6.

Ziemer, A. The controversy: HDTV via glass fibre or satellite?
Funkschau (West Germany); 1987 May 22; (11): 34-7.

Ziemer, A. HDTV: 'We need a world standard'. Funkschau (West
Germany); 1987 December 18; (26): 29-31.

Ziemer, A. High definition TV-electronic picture production
method of the future? Fernseh und Kino-Technik (West
Germany); 1985 November; 39(11): 521-2.

Zimmer, G.; Weinerth, H. Key technologies in microelectronics.
12. Applications in innovative highly complex systems.
Elektronik (West Germany); 1989 May 26; 38(11): 96-107.

Zschunke, W.; Suto, K.; Yamashita, I. Optical fiber transmission
of high definition television signals by analog intensity
modulation. Transactions of the Institute of Electronic &
Communication Engineers of Japan Section E (Japan); 1985
March; E68(3): 188-94.

Zwilling, H. Satellite broadcasting and cable television in and
outside Europe. NTZ Nachrichtentechnische Zeitschrift
(Germany); 1983 June; 36(6): 372-7.

Part Two: Papers and Conference Proceedings

Agresti, M. Some experiments of motion picture production using HDTV. In: Electro/87 and Mini/Micro Northeast: Focusing on the OEM. Conference Record. Los Angeles: Electronic Conventions Management; 1987: 33. 1. 1-33. 0. 3.

Ahmed, S.N. Future technical evolution of national spectrum management. In: SPECTRUM 20/20. A Symposium on Spectrum Usage: Future Directions in Canada. Symposium Proceedings. Ottawa, Ontario, Canada: Radio Advisory Board Canada; 1987; III/4: 1-16.

Akiyama, Y. et al. A 1280*980 pixel CCD image sensor. In: Winner, L. 1986 IEEE International Solid-State Circuits Conference. Digest of Technical Papers. Coral Gables, FL: Lewis Winner; 1986: 96-7.

Ando, K. et al. A 54-inch (5:3) high-contrast and brightness rear-projection display for high-definition TV. In: Morreale, J. 1985 SID International Symposium. Digest of Technical Papers. New York: Pallisades Institute of Research Services; c1985: 274-6.

Ando, K. et al. A 54-inch (5:3) high-contrast high-brightness rear-projection display for high-definition TV. In: Society for Information Display 1985 International Symposium, Seminar and Exhibition -- SID '85. Playa del Ray, CA: Society for Information Display; 1985.

Annegarn, M.J.J.C. The counting of received pixels in TV receiving systems. In: IEEE 1988 International Conference on Consumer Electronics. Digest of Technical Papers. New York: IEEE; 1988: 194-5.

Annegarn, M.J.J.C.; Arragon, J.-P.; Jackson, R.N. High definition MAC: the compatible route to HDTV. In: IBC 86. International Broadcasting Convention. London: IEE; 1986: 153-7.

Annegarn, M.J.J.C.; Arragon, J.-P.; Jackson, R.N. High definition MAC: the compatible route to HDTV. In: Cantraine, G.; Destine, J. New Systems and Services in Telecommunications, III: Networks, Cables, Satellites - The What, the How, the Why: Proceedings of the Third International Conference. Amsterdam: North-Holland; 1987: 321-5.

Aono, K.; Toyokura, M.; Araki, T. A 30 ns (600 MOPS) image processor with a reconfigurable pipeline architecture. In: Procedings of the IEEE 1989 Custom Integrated Circuits Conference. New York: IEEE; 1989: 24.4/1-4.

Apple, G.G.; Tsou, H.E. Data compression for high definition TV: an NTSC compatible approach. In: National Telesystems Conference. NTC '82. Systems for the Eighties. New York: IEEE; 1982; El.3: 1-4.

Armbruster, H. Broadband ISDN-the network of the future: applications and compliance with user requirements. In: GLOBECOM '86: IEEE Global Telecommunications Conference. Communications Broadening Technology Horizons. Conference Record. New York: IEEE; 1986; 1: 484-90.

Armbruster, H. Future communications with broadband ISDN: services, applications and technical implementation. In: Mastroddi, F. Electronic Publishing: The New Way to Communicate. Proceedings of the Symposium. London: Kogan Page; 1987: 191-203.

Arragon, J.P. et al. Instrumentation of a compatible HD-MAC coding system using DATV. In: IBC 1988: International Broadcasting Convention. London: IEE; 1988: 57-61.

Arragon, J.P.; Fonsalas. F.; Haghiri, M. Motion compensated interpolation techniques for HD-MAC. In: IBC 1988: International Broadcasting Convention. London: IEE; 1988: 70-3.

Baack, C.; Heydt, G.; Walf, G. Broadband distribution techniques for future broadband communications networks. In: TELE MATICA 88: Internationaler Fachkongress fur Integrierte Telekommunikation, Telematik, Kabel-und

Satellitenkommunikation (TELEMATICA 88: International Conference for Integrated Telecommunications, Telematics, Cable and Satellite Communications). Munich, West Germany: Verlag Reinhard Fischer; 1988: 160-7.

Baack, C.; Heydt, G.; Vathke, J. Digital integrated services broadband network with HDTV-capability. In: ISS '84. XI International Switching Symposium. Milan, Italy: Association Elettrotecnica & Elettronica Italiana; 1984; 3: 32C/1-6.

Baack, C.; Heydt, G.; Walf, G. TV/HDTV distribution by optical fiber systems. In: Optical Fiber Communication Conference: Summaries of Papers. Piscataway, NJ: IEEE; 1989: 56.

Bailey, W.H. Cable industry perspective on HDTV delivery. In: IEEE EASCON '88. 21st Annual Electronics and Aerospace Conference: How will Space and Terrestrial Systems Share the Future? Conference Proceedings. New York: IEEE; 1988: 185-6.

Bernard, P.; Veillard, M. Analysis of spatio-temporal sub-sampling structures of an HDTV signal for transmission over a MAC channel. In: Third International Colloquium on Advanced Television Systems: HDTV '87. Colloquium Proceedings. Montreal, Quebec, Canada: CBC; 1988; 1: 6/2/1- 28.

Besier, H. et al. An experimental system for wideband services on fibre optic subscriber lines using frequency multiplex. In: Cantraine, G.; Destine, J. New Systems and Services in Telecommunications, III: Networks, Cables, Satellites - The What, the How, the Why? Proceedings of the Third International Conference. Amsterdam: North-Holland; 1987: 255-60.

Billotet-Hoffmann, C.; Sauerburger, H. Problems of fieldrate conversions in HDTV. In: Procedings of SPIE - The International Society for Optical Engineering. Bellingham, WA: SPIE; 1986; 594: 49-56.

Binns, J.F.H. Television transmitters-the modulation process. In: International Conference on the History of Television - From Early Days to the Present. London: IEE; 1986: 126-30.

Blume, G. et al. Optical cable broadband subscriber network. In: Proceedings of 37th International Wire and Cable Sympo sium. Fort Monmouth, NJ: US Army Communications-

Electronics Command; 1988: 122-8.

Boegels, P.W. The EUREKA HDTV project-philosophy and practice. In: IBC 1988: International Broadcasting Convention. London: IEE; 1988: 430-7.

Boerner, R. Progress in projection of parallax-panoramagrams onto wide-angle lenticular screens. In: Proceedings of SPIE - The International Society for Optical Engineering. Bellingham, WA: SPIE; 1987; 761: 35-43.

Bolle, G. The future of television technology. In: TELEMATICA 88: Internationaler Fachkongress fur Integrierte Telekommunikation, Telematik, Kabel-und Satellitenkommunication (TEL-MATICA 88: International Conference for Integrated Tele-communications, Telematics, Cable and Satellite Communications). Munich, West Germany: Verlag Reinhard Fischer; 1988: 555-66.

Bourguignat, E. Complementarity of the psychophysical and subjective approaches in HDTV. In: Third International Colloquium on Advanced Television Systems: HDTV '87. Colloquium Proceedings. Montreal, Quebec, Canada: CBC; 1988; 1: 1/4/1-16.

Bourguignat, E. Psychovisual approach of HDTV bit rate reduction. In: IEE Colloquium on 'HDTV Bandwidth Reduction'. London: IEE; 1987; c2; 24: 1-5.

Boyer, R.; Melwig, R. Evolutionary approach and compatible HDTV system in Europe. In: Third International Colloquium on Advanced Television Systems: HDTV '87. Colloquium Proceedings. Montreal, Quebec, Canada: CBC; 1988; 1: 5/2/1- 18.

Bush, S.; Cripps, D. HDTV Conference 1989. Proceedings of the First Annual Conference on High Definition Television (New York, June 1989). Westport, CT: Meckler; 1989.

Bush, S.; Cripps, D. HDTV International 1989. Proceedings of the First Annual International Conference on High Definition Television (London, September 1989). London: Meckler; 1989.

Byrd, J.C. New video standards. In: Proceedings of the IEEE 1983 National Aerospace and Electronics Conference. NAECON 1983. New York: IEEE; 1983; 1: 322-6.

Carpenter, J.B. ISDN - transition to broadband access. In: Proceedings of the National Electronics Conference. USA:

Professional Educational International, Incorporated; 1986; 40 part 1: 544-46.

Chatel, J. Towards a world studio standard for high definition television. In: IBC 1988: International Broadcasting Convention. London: IEE; 1988: 8-11.

Chen, S.C. et al. Colorimetry and constant luminance coding in a compatible HDMAC system. In: IBC 1988: International Broadcasting Convention. London: IEE; 1988: 45-8.

Chiarglione, L.; Guglielmo, M.; van Veen, W.M.D. A family of frame structures for local digital video distribution. In: GLOBECOM '85. IEEE Global Communications Conference. Conference Record. Communication Technology to Provide New Services. New York: IEEE; 1985; 3: 1340-4.

Childs, I. DATV techniques for HDTV display upconversion and transmission coding. In: IEE Colloquium on 'HDTV Band-width Reduction'. London: IEE; 1987; c6; 24: 1-5.

Childs, I. Introduction to high definition television. In: Video, Audio and Data Recording. Sixth International Conference. London: IERE; 1986: 75-80.

Chin, D. et al. The Princeton Engine: a real-time video system simulator. In: IEEE 1988 International Conference on Consumer Electronics. Digest of Technical Papers. New York: IEEE; 1988: 144-5.

Choquet, B.; Pele, D. Extraction of the spatio-temporal parameters from a high definition image sequence. In: Third International Colloquium on Advanced Television Systems: HDTV '87. Colloquium Proceedings. Montreal, Quebec, Canada: CBC; 1988; 1: 6/3/1-6.

Choquet, B.; Siohan, P. Enhancement techniques of a motion detector in high definition television. In: Second International Conference on Image Processing and its Applications. London: IEE; 1986: 220-3.

Chouinard, G. Broadcasting of HDTV: beyond the vision signal. In: Third International Colloquium on Advanced Television Systems: HDTV '87. Colloquium Proceedings. Montreal, Quebec, Canada: CBC; 1988; 1: 4/7/1-23.

Chu, S.; Chen., T.C. Flexible format video sequence processing simulation system. In: Proceedings of SPIE - The International Society for Optical Engineering. Bellingham,

WA: SPIE; 1986; 707: 116-123.

Clemow, R.D. The use of simulation for HDTV bandwidth reduction studies. In: IEE Colloquium on 'HDTV Bandwidth Reduction'. London: IEE; 1987; c3; 24: 1-5.

Craig, D.; Craig, M. Video fundamentals. In: NCGA '89 Conference Proceedings. 10th Annual Conference and Exposition Dedicated to Computer Graphics. Fairfax, VA: National Computer Graphics Association; 1989; 3: 416-36.

Crowther, G.O. C, D, D/sub 2/ MAC decoder architecture. In: Cantraine, G.; Destine, J. New Systems and Services in Telecommunications, III: Networks, Cables, Satellites - The What, the How, the Why? Proceedings of the Third International Conference. Amsterdam: North-Holland; 1987: 335-9.

Crutchfield, E.B. Broadcasting high definition television. In: IBC 1988: International Broadcasting Convention. London: IEE; 1988: 34-6.

Cutts, D.J. Subscription management in satellite TV services in Europe-structure and objectives. In: IBC 1988: International Broadcasting Convention. London: IEE; 1988: 333-5.

Dalton, C.J. Video standards review. In: IEE Colloquium on 'Video Standards and their Interconnection'. London: IEE; 1989: 2/1-9.

David, D. et al. INP 20. An image neighborhood processor for large kernels. In: Proceedings of IAPR Workshop on Computer Vision: Special Hardware and Industrial Applications. Tokyo: University of Tokyo; 1988: 241-4.

Davies, K.P.; Field, K.R.; Sawyer, B. HDTV services - a progress report. In: Canadian Satellite User Conference, 1987 - Conference Proceedings. Ottowa, Canada: Telesat; 1987: 335.

de Haan, G.; Crooijmans, W. Subsampling techniques for high definition MAC. In: IBC 86. International Broadcasting Convention. London: IEE; 1986: 158-62.

DePriest, G.L.; Schmidt, G.M. Advanced television: a terrestrial perspective. In: IBC 1988: International Broadcasting Convention. London: IEE; 1988: 438-40.

Dobbie, W.H. The D2-SMAC system for bandwidth efficient FM TV transmission. In: Third International Colloquium on Advanced

Television Systems: HDTV '87. Colloquium Proceedings. Montreal, Quebec, Canada: CBC; 1988; 1: 3/2/1-11.

Doyle, T.; van Alphen, W.M.; Vriens, L. HDTV: a display perspective. In: IBC 1988: International Broadcasting Convention. London: IEE; 1988: 212-15.

Drury, G.M.; Carmen, P.R.; Sparks, M.B. Recording enhanced definition television: a wideband analogue component approach. In: Video, Audio and Data Recording. Sixth International Conference. London: IERE; 1986: 87-94.

Duvic, G.; Veillard, J. Description of a D2-HD-MAC/packet chain. In: IBC 1988: International Broadcasting Convention. London: IEE; 1988: 12-16.

Dwyer, J.M. Future telecommunications developments within the cost framework. In: 7th European Conference on Electrotechnics: Advanced Technologies and Processes in Communication and Power Systems - EUROCON 86. Paris: Comite EUROCON 86; 1986: 322-29. Eouzan, J.Y.; Boyer, R. A progressive scanning 1250/501 HDTVcolour camera and processing based on quincunx sampling. In: IBC 1988: International Broadcasting Convention. London: IEE; 1988: 174-6.

Eto, Y. et al. A practical approach for achieving an HDTV digital VTR. In: Third International Colloquium on Advanced Television Systems: HDTV '87. Colloquium Proceedings. Montreal, Quebec, Canada: CBC; 1988; 1: 2/6/1-11.

Fernandez, A. et al. A raster assembly processor (RAP) for integrated HDTV display of video and image windows. In: GLOBECOM Tokyo '87. IEEE/IECE Global Telecommunications Conference 1987. Conference Record. New York: IEEE; 1987; 2: 731-9.

Fernando, G.M.X. Motion compensated field rate conversion for HDTV display. In: IEE Colloquium on 'Motion Compensated Image Processing'. London: IEE; 1987; c5; 11: 1-4.

Fernando, G.M.X.; Parker, D.W. Motion compensated field rate conversion for HDTV display. In: Third International Colloquium on Advanced Television Systems: HDTV '87. Colloquium Proceedings. Montreal, Quebec, Canada: CBC; 1988; 1: 6/5/1-18.

Fernando, G.M.X.; Parker, D.W. Motion compensated field rate conversion for HD-MAC display. In: IBC 1988: International

Broadcasting Convention. London: IEE; 1988: 216-19.

Field, K.R.; Galt, J. Experiences with HDTV drama production. In: Third International Colloquium on Advanced Television Systems: HDTV '87. Colloquium Proceedings. Montreal, Quebec, Canada: CBC; 1988: 33. 3. 1-33. 3.4.

Field, K.R.; Galt, J. First experiences with drama production in HDTV. In: Electro/87 and Mini/Micro Northeast: Focusing on the OEM. Conference Record. Los Angeles: Electronic Conventions Management; 1987; c3; 33: 1-4.

Flaherty, J.; O'Connor, R.A.; Ramasastry, J. High definition television service in the 12 GHz band. In: AIAA 9th Communications Satellite Systems Conference. New York: AIAA; 1982.

Foisel, H.-M. Ten-channel coherent HDTV/TV distribution system. In: ECOC 87: 13th European Conference on Optical Communication. Technical Digest. Helsinki, Finland: Consulting Committee Professional Electroengineering Organization of Finland; 1987; 1: 287-90.

Fonsalas, F.; Lejard, J.Y. A method for chrominance contour enhancement applied to HD-MAC television pictures. In: IBC 86. International Broadcasting Convention. London: IEE; 1986: 225-8.

Forrest, J.R. High definition television: when do we start in Europe. In: Schuringa, T.M. EuroComm 88: Proceedings of the International Congress on Business, Public and Home Communications. Amsterdam: North-Holland; 1988: 9-20.

Fourdeux, H.; Claverie, C. Digital HDTV signal transmission on optical fibre. In: IBC 1988: International Broadcasting Convention. London: IEE; 1988: 280-3.

Free, L.R. An antipodean view of higher definition television. In: IBC 1988: International Broadcasting Convention. London: IEE; 1988: 25-9.

Freeman, K.G. et al. Experimental work towards high fidelity television. In: IBC 82. International Broadcasting Convention. London: IEE; 1982: 140-3.

Fujio, T. et al. High-definition television system-signal standards and transmission. In: International Broadcasting Convention. London: IEE; 1980: 14-18.

Fujio, T.; Sugimoto, M.; Masuko, Y. Development of high

definition television technique in Japan. In: Cantraine, G.; Destine, J. New Systems and Services in Telecommunications II. Proceedings of the Second International Conference. Amsterdam: North-Holland; 1984: 369-77.

Fujita, Y. et al. High-definition television evaluation for remote handling task performance. In: Proceedings of the 34th Conference on Remote Systems Technology. La Grange Park, IL: ANS; 1987: 59-68.

Fujita, Y. et al. High-definition television: a new influence on data display. In: American Nuclear Society Winter Meeting. La Grange Park, IL: American Nuclear Society; 1986.

Fukinuki, T.; Hirano, Y.; Yoshigi, H. Extended definition television-higher quality image with compression technology. In: GLOBECOM Tokyo '87. IEEE/IECE Global Telecommunications Conference 1987. Conference Record. New York: IEEE; 1987; 1: 400-4.

Fukinuki, T.; Hirano, Y.; Yoshigi, H. NTSC-full-compatible extended-definition TV-proto model and motion adaptive processing. In: IEEE Global Telecommunications Conference--GLOBECOM '85. Piscataway, NJ: IEEE; 1985.

Gaggioni, H.; Robbins, J. A flexible memory system for the integration of multiple studio quality video and image windows on an HDTV display. In: Third International Colloquium on Advanced Television systems: HDTV '87. Colloquium Proceedings. Montreal, Quebec, Canada: CBC; 1988; 1.

Gagnon, D. et al. The influence of resolution and content on subjective evaluations on picture quality: a study series. In: Third International colloquium on Advanced Television Systems: HDTV '87. Colloquium Proceedings. Montreal, Quebec, Canada: CBC; 1988; 1: 1/3/1-17.

Gardiner, P. The D-MAC transmission standard. In: Direct Broadcast by Satellite Conference. London: Consert; 1988: 61-71.

Gerber, J. A relation between H/sub O/(z) in deflection yokes and its influence on convergence errors. In: Morreale, J. 1985 SID International Symposium. Digest of Technical Papers. New York: Pallisades Institute of Research Services; 1985: 174-7.

Gerhard-Multhaupt, R.; Tepe, R. Two-dimensional spatial light modulators for high-resolution TV applications. In: Electro-optic and Photorefractive Materials, Proceedings of the International School on Material Science and Technology. Berlin and New York: Springer-Verlag; 1987: 377-80.

Gerrard, G.A. A review of the applications of different types of television codecs. In: Third International Conference on Telecommunication Transmission. London: IEEE; 1985: 282-5.

Glenn, W.E. High-definition television. In: Morreale, J. Society for Information Display 1988 Seminar Lecture Notes. Playa del Rey, CA: SID; 1988; 1: 4/1-27.

Glenn, W.E. Solid-state-driven deformable television light modulator. In: Morreale, J. 1987 SID International Symposium. Digest of Technical Papers. First Edition. New York: Palisades Institute Research Services; 1987: 72-4.

Glenn, W.E.; Glenn, K.G. Improved HDTV with compatible transmission. In: Third International Colloquium on Advanced Television Systems: HDTV '87. Colloquium Proceedings. Montreal, Quebec, Canada: CBC; 1988; 1: 4/5/1-12.

Golding, L.S. Advanced television signal transmission via satellite. In: Electronic Imaging '88: International Electronic Imaging Exposition and Conference. Advance Printing of Paper Summaries. Waltham, MA: Institute of Graphic Communications; 1988; 1: 84-8.

Green, N.W. HDTV signal origination. In: IEE Colloquium on 'Progress Towards HDTV'. London: IEE; 1989: 1-1/4.

Haag, H.-G.; Kuchenbecker, H.-P. Broadcast distribution service using glass-fibres in a future integrated communications network. In: TELEMATICA 88: Internationaler Fachkongress fur Integrierte Telekommunikation, Telematik, Kabel-und Satellitenkommunikation (TELEMATICA 88: International Conference for Integrated Telecommunications, Telematics, Cable and Satellite Communications). Munich, West Germany: Verlag Reinhard Fischer; 1988: 432-9.

Habermann, W. Stereoscopic television: an alternative to and an extension of HDTV. In: Cantraine, G.; Destine, J. New Systems and Services in Telecommunications II. Proceedings of the Second International Conference. Amsterdam: North-

Holland; 1984: 41-14.

Haddon, N.J. A high resolution electronic art system for the print me dia. In: Electro/87 and Mini/Micro Northeast: Focusing on the OEM. Conference Record. Los Angeles: Electronic Conventions Management; 1987: 33. 4. 1-33. 4. 3.

Haskell, B.G. High definition television (HDTV)-compatibility and distribution. In: Globecom '83. IEEE Global Telecommunications Conference. Conference Record. New York: IEEE; 1983; 2: 1070-5.

Haskell, B.G. Semi-compatible high definition television using field differential signals. In: DeWilde, P.; May, C.A. Links for the Future. Science, Systems and Services for Communications. Proceedings of the International Conference on Communications-ICC 84. Amsterdam: North-Holland; 1984; 1: 491-4.

Haskell, B.G. Semi-compatible high definition television using field differential signals. In: ICC '84 Conference Record. Amsterdam: Elsevier Science Publishers; 1984.

Haskell, P.; Tzou, K.-H.; Hsing, T.R. A lapped-orthogonal-transform based variable bit-rate video coder for packet networks. In: ICASSP-89: 1989 International Conference on Acoustics, Speech and Signal Processing. New York: IEEE; 1989; 3: 1905-8.

Hayano, S. et al. A GaAs 8*8 matrix switch LSI for high-speed digital communications. In: GaAs IC Symposium: IEEE Gallium Arsenide Integrated Circuit Symposium. Technical Digest. New York: IEEE; 1987: 245-8.

Hewitt, C.C. Advanced television systems and the home satellite television market. In: IEEE EASCON '88. 21st Annual Electronics and Aerospace Conference: How will Space and Terrestrial Systems Share the Future? Conference Proceedings. New York: IEEE; 1988: 181-4.

Himeno, A. et al. Experimental optical switching system using space-division matrix switches gated by laser diodes. In: GLOBECOM '88. IEEE Global Telecommunications Confer ence and Exhibition - Communications for the Information Age. Conference Record. New York: IEEE; 1988; 2: 928-32.

Hofker, U.; Teich, G. A digital fiber optic HDTV transmission system in Berlin. In: Third International Colloquium on Advanced Television Systems: HDTV '87. Colloquium

Proceedings. Montreal, Quebec, Canada: CBC; 1988; 1: 3/1/1- 21.

Hopkins, R. HDTV, past and present. In: Electro/87 and Mini/Micro Northeast: Focusing on the OEM. Conference Record. Los Angeles: Electronic Conventions Management; 1987: 33. 0. 1-33. 0. 3.

Horstman, R.A. Videodisc and player for HDMAC. In: IBC 1988: International Broadcasting Convention. London: IEE; 1988: 224-7.

Hurault, J.-P.; Arragon, J.-P. Development of advanced HD-MAC coding algorithms. In: IBC 1988: International Broadcasting Convention. London: IEE; 1988: 54-6.

Iesaka, M. et al. Analysis of charge transfer loss in dual read-out registers used for HDTV CCD image sensor. In: Extended Abstracts of the 20th (1988 International) Conference on Solid State Devices and Materials. Tokyo: Business Center for the Academic Society of Japan; 1988: 359-62.

Ishibashi, S. The HMC-1000 high definition multi-scan still camera system. In: Electronic Imaging '88: International Electronic Imaging Exposition and Conference. Advance Printing of Paper Summaries. Waltham, MA: Institute of Graphic Communications; 1988; 1: 114-19.

Isnardi, M.A. et al. A single channel, NTSC compatible widescreen EDTV system. In: Third International Colloquium on Advanced Television Systems: HDTV '87. Colloquium Proceedings. Montreal, Quebec, Canada: CBC; 1988; 2.

Isozaki, Y.; Ogusu, C.; Kumada, J. Pick up tube and camera for HD-TV. In: IBC 82. International Broadcasting Convention. London: IEE; 1982: 152-4.

Jackson, R.N. High defintion television: a new influence on data display. In: EUROGRAPHICS '87. Amsterdam: Elsevier Science Publishers; 1987.

Jackson, R.N.; Tan, S.L. System concepts in high fidelity television. In: IBC 82. International Broadcasting Convention. London: IEE; 1982: 135-8.

James, I.J.P. Some aspects of colour television at EMI. In: International Conference on the History of Television - From Early Days to the Present. London: IEE; 1986: 47.

Johnson, C. Europe at the crossroads: D/D2-MAC-the satellite

television standard. In: TELEMATICA 88: Internationaler
Fachkongress fur Integrierte Telekommunikation, Telematik,
Kabel-und Satellitenkommunikation (TELEMATICA 88:
International Conference for Integrated Telecommunications,
Telematics, Cable and Satellite Communications). Munich,
West Germany: Verlag Reinhard Fischer; 1988: 514-20.

Kauff, P.; Schafer, R. Comparison of sequential and interlaced
scan for HDTV with regard to spatial and temporal
resolution. In: IEEE 1988 International Conference on
Consumer Electronics. Digest of Technical Papers. New York
: IEEE; 1988: 62-3.

Kirby, R.C.; Sturzak, R.G. The international dimension in
spectrum utilization. In: SPECTRUM 20/20. A Symposium
on Spectrum Usage: Future Directions in Canada. Symposium
Proceedings. Ottawa, Ontario, Canada: Radio Advisory Board; 1987;
III/1: 1-11.

Kishimoto, R. et al. The development of broadband transmission
technologies in NTT. In: Third International Conference on
Telecommunication Transmission. London: IEEE; 1985:
11-14.

Kishimoto, R.; Sakurai, N. A high-definition TV transmission
system using adaptive subsampling. in: GLOBECOM
Tokyo '87. IEEE/IECE Global Telecommunications Conference 1987. New York: IEEE; 1987; 1: 411-15.

Kishimoto, T.; Sakurai, N. A high-definition TV transmission
system using adaptive subsampling. In: GLOBECOM
Tokyo '87. IEEE/IECE Global Telecommunications
Conference 1987.
Conference Record. New York: IEEE; 1987; 1: 411-15.

Klemmer, W. HDTV camera technology: KCH 1000-a multi-standard
HDTV camera system. In: TELEMATICA 88: Internationaler
Fachkongress fur Integrierte Telekommunikation, Telematik,
Kabel-und Satellitenkommunikation (TELEMATICA 88:
International Conference for Integrated Telecommunications,
Telematics, Cable and Satellite Communications. Munich,
West Germany: Verlag Reinhard Fischer; 1988: 588-97.

Klemmer, W. The KCH 1000 a multi-standard HDTV camera system.
In: IBC 1988: International Broadcasting Convention. London: IEE; 1988: 177-80.

Klemmer, W.H. The color camera in the HDTV studio-an innovative concept. In: Third International Colloquium on Advanced Television Systems: HDTV '87. Colloquium Proceedings. Montreal, Quebec, Canada: CBC; 1988; 1: 2/4/1-16.

Kondo, M. et al. 32 switch-elements integrated low-crosstalk LiNbO/sub 3/ 4*4 optical matrix switch. In: IOOC-ECOC '85. 5th International Conference on Integrated Optics and Optical Fibre Communication and 11th European Conference on Optical Communication. Technical Digest. Genova, Italy: Instituto Int. Comunicazioni; 1985; 1: 361-4.

Kou-Hu T. et al. Compatible HDTV coding for broadband ISDN. In: GLOBECOM '88. IEEE Global Telecommunications Conference and Exhibition - Communications for the Information Age. Conference Record. New York: IEEE; 1988; 2: 743-9.

Kunt, M. ed.; Huang, T.S. ed. Image coding. In: Proceedings of SPIE - The International Society for Optical Engineering. Bellingham, WA: SPIE; 1986; 594: 1-351.

Lambert, D.T.; Senior, J.M.; Faulkner, D.W. High definition television transmission on single mode fibre for the local loop. In: IEE Colloquium on 'Fibre Optic LANS and Techniques for the Local Loop'. London: IEE; 1989: 9/1-6.

Lechner, B.J. High definition TV. In: Morreale, J. 1985 SID International Symposium. Digest of Technical Papers. New York: Pallisades Institute for Research Services; 1985: 14.

Lechner, B.J. High definition TV. In: Society for Information Display 1985 International Symposium, Seminar and Exhibition --SID '85. Playa del Ray, CA: Society for Information Display; 1985.

Lewis, R.G.R. et al. A second generation HDTV camera. In: IBC 1988: International Broadcasting Convention. London: IEE; 1988: 170-3.

Lo, C.N.; Smoot, L.S. Integrated fiber optic transmission of FM HDTV and 622 Mb/s data. In: IEEE 1989 MTT-S International Microwave Symposium Digest. New York: IEEE; 1989; 2: 703-4.

LoCicero, J.L.; Pazarci, M.; Rzeszewski, T.S. Aspect ratio improvements in a compatible HDTV system. In: IEEE International Conference on Communications 1985. New York: IEEE; 1985; 1: 423-7.

LoCicero, J.L.; Pazarci, M.; Rzeszewski, T.S. Edge stitching of a wide-respect ratio HDTV image. In: IEEE International Conference on Communications '86. ICC '86: 'Integrating the World Through Communications'. Conference Record. New York: IEEE; 1986; 1: 436-40.

LoCicero, J.L.; Pazarci, M.; Rzeszewski, T.S. Image reconstruction in a wide-aspect ratio HDTV system. In: IEEE International Conference on Communications--ICC '86. Piscataway, NJ: IEEE; 1986.

Lodge, N.K. Bit-rate reduction techniques for HDTV transmission links. In: IEE Colloquium on 'HDTV Bandwidth Reduction'. London: IEE; 1987; c5; 24: 1-5.

Lothian, J.S.; Beech, B. H. Transmission using MAC signal formats. In: IEE Colloquium on 'Progress Towards HDTV'. London: IEE; 1989: 3/1-6.

Lucas, K. B-MAC and HDTV-does it fit? In: Third International Colloquium on Advanced Television Systems: HDTV '87. Colloquium Proceedings. Montreal, Quebec, Canada: CBC; 1988; 1: 4/3/1-20.

Lupker, S.J.; Hearty, P.J. Analysing the subjective effects of multiple sources of impairment in television signals. In: Third International Colloquium on Advanced Television Systems: HDTV '87. Colloquium Proceedings. Montreal, Quebec, Canada: CBC; 1988; 1: 1/2/1-17.

Lyner, A.G. Design of HDTV systems: transmission channel noise. In: Colloquium on Noise in Images. London: IEE; 1987; c17; 3: 1-3, 5.

Maddern, T.S. Analysis of multi-slot connections. In: Second IEE National Conference on Telecommunications. London: IEE; 1989; c300: 321-326.

Mahler, G. The state of the large-picture HDTV. In: TELEMATICA 88: Internationaler Fachkongress fur Integrierte Telekommunikation, Telematik, Kabel-und Satellitenkommunikation (TELEMATICA 88: International conference for Integrated Telecommunications, Telematics, Cable and Satellite Communications). Munich, West Germany: Verlag Reinhard Fischer; 1988: 598-604.

Makitalo, O. Broadcasting of HDTV via satellite. In: IEEE International Conference on Communications '88: Digital

Technology - Spanning the Universe. Conference Record.
New York: IEEE; 1988; 1: 152-6.

Manabe, S. et al. A 2-million pixel CCd imager overlaid with an
amorphous silicon photoconversion layer. In: Winner, L. 1988
IEEE International Solid-State Circuits Conference. Digest
of Technical Papers. 31st ISSCC. First Edition. Coral
Gables, FL: Lewis Winner; 1988 February: 50-1, 297.

Mano, Y. et al. MUSE video disc. In: Third International
Colloquium on Advanced Television Systems: HDTV '87.
Colloquium Proceedings. Montreal, Quebec, Canada: CBC;
1988; 1: 4/8/1-10.

Mayo, B.J. Early development of electron guns and cathode-ray
tubes for television. In: International Conference on the
History of Television - From Early Days to the Present.
London: IEE; 1986: 69-72.

McKnight, L.; Neil, S. The HDTV war: the politics of HDTV
standardization. In: Third International Colloquium on
Advanced Television Systems: HDTV '87. Colloquium
Proceedings. Montreal, Quebec, Canada: CBC; 1988; 1: 5/6/
1-17.

McMann, R.H.; Goldberg, A.A.; Rossi, J.P. A two channel
compatible HDTV broadcast system. In: Cantraine, G.;
Destine, J. New Systems and Services in Telecommunications
II. Proceedings of the Second International Conference.
Amsterdam: North-Holland; 1984: 379-85.

Melwig, R. Colorimetry in HDTV: up-to-date solutions for a new
system. In: Proceedings of SPIE - The International Society
for Optical Engineering. Bellingham, WA: SPIE; 1986; 594:
41-48.

Minami, F.; Sakurai, M.; Ninomiya, Y. MUSE decoder. In: IEEE
1988 International Conference on Consumer Electronics. Digest of Technical Papers. New York: IEEE; 1988: 192-3.

Minzer, S.E.; Spears, D.R. Principles for defining a signalling
protocol for complex calls in an ISDN. In: IEEE
International Conference on Communications '87:
Communications-Sound to Light. Proceedings. New York:
IEEE; 1987; 1: 314-18.

Miyagawa, R. et al. A new, pre-discharge boron doping method for
amorphous silicon photoconversion layer in HDTV image

sensor. In: International Electron Devices Meeting. Technical Digest. New York: IEEE; 1988: 74-7.

Morcom, R. Investigation of HDTV parameters using an image processor. In: Third International Colloquium on Advanced Television Systems: HDTV '87. Colloquium Proceedings. Montreal, Quebec, Canada: CBC; 1988; 1.

Murakami, H. et al. An 8-inch diagonal pulse discharge panel with internal memory for a color TV display. In: 1984 SID international symposium. Digest of technical papers. New York: Palisades Institute for Research Services; 1984: 87- 90.

Nadan, J.S. Frame stores and advanced signal processing in future television receivers. In: ELECTRO/85. Conference record. Los Angeles: Electronic Conventions Management; 1985; 13/2: 1-3.

Nakamura, Y. HDTV-past and present. In: Third International Colloquium on Advanced Television Systems: HDTV '87. Colloquium Proceedings. Montreal, Quebec, Canada: CBC; 1988; 1: 0/1/1-9.

Ninomiya, Y. et al. HDTV broadcasting and transmission system-MUSE. In: Third International Colloquium on Advanced Television Systems: HDTV '87. Colloquium Proceedings. Montreal, Quebec, Canada: CBC; 1988; 1: 4/1/1-31.

Ninomiya, Y. et al. A motion vector detector for MUSE encoder. In: IEEE International Conference on Communications '86. ICC '86: 'Integrating the World Through Communications'. Conference Record. New York: IEEE; 1986; 2: 1280-4.

Nishizawa, T. et al. HDTV and ADTV transmission systems-MUSE and its family. In: IBC 1988: International Broadcasting Convention. London: IEE; 1988: 37-40.

Nishizawa, T.; Tanaka, Y. Standards conversion with HDTV. In: Third International Colloquium on Advanced Television Systems: HDTV '87. Colloquium Proceedings. Montreal, Quebec, Canada: CBC; 1988; 1: 2/3/1-19.

Nosu, K.; Toba, H. Performance consideration on high capacity optical FDM networks. In: GLOBECOM '88. IEEE Global Telecommunications Conference and Exhibition - Communications for the Information Age. Conference Record. New York: IEEE; 1988; 1: 496-500.

Oakley, K.A. An economic way to see in the broadband dawn (passive optical network). In: GLOBECOM '88. IEEE Global Telecommunications Conference and Exhibition - Communications for the Information Age. Conference Record. New York: IEEE; 1988; 3: 1574-8.

Ohmura, T.; Tanaka, Y.; Kurita, T. The development of an HDTV to PAL standards converter and its picture quality. In: IBC 86. International Broadcasting Convention. London: IEE; 1986: 221-4.

Okai, H. Towards the realization of HDTV-situation in Japan. In: Third International Colloquium on Advanced Television Systems: HDTV '87. Colloquium Proceedings. Montreal, Quebec, Canada: CBC; 1988; 1: 5/4/1-9.

Okazaki, T. et al. Implementation of a HDTV codec using a hybrid quantizer. In: GLOBECOM Tokyo '87. IEEE/IECE Global Telecommunications Conference 1987. Conference Record. New York: IEEE; 1987; 1: 421-5.

Oliphant, A. A multiplexed routing system for a digital TV studio centre. I. The advantages of a multiplexed system. In: IEE Colloquium on 'TV Studios from A/D'. London: IEE; 1987; c5; 11: 1-3.

Olshansky, R.; Lanzisera, V.A. Sixty-channel FM video subcarrier multiplexed optical communication system. In: Optical Fiber Communication Conference, 1988 Technical Digest Series, Summaries of Papers. Washington, DC: Optical Society of America; 1988; 1: 192.

Olshansky, R.; Lanzisera, V. Subcarrier multiplexed passive optical nertwork for low-cost video distribution. In: Optical Fiber Communication Conference: Summaries of Papers. Piscataway, NJ: IEEE; 1989: 57.

Parker, D.W. HDTV bandwidth compression: the compatible approach. In: IEE Colloquium on 'HDTV Bandwidth Reduction'. London: IEE; 1987; c1; 24: 1-5.

Parker, D.W. Receivers and displays for HDTV. In: IEE Colloquium on 'Progress Towards HDTV'. London: IEE; 1989: 4/1-4.

Pauchon, B. HDTV-summary of operational and eocnomic features with a view to international standardization. In: IBC 1988: International Broadcasting Convention. London: IEE; 1988: 17-20.

Paulson, C.R. Laser and fiber optic technologies: lighting the way for television communication growth. In: Electronic Imaging '88: International Electronic Imaging Exposition and Conference. Advance Printing of Paper Summaries. Waltham MA: Institute of Graphic Communication; 1988; 1: 112-13.

Pele, D.; Choquet, B. Estimation and segmentation of apparent motion fields in HDMAC. In: IBC 1988: International Broadcasting Convention. London: IEE; 1988: 74-7.

Phillips, G.J. Future developments (satellite broadcasting). In: Colloquium on Satellite Broadcasting. London: IEE; 1980: 32 pp.

Pica, A.P. On towards the future and high definition television. In: ELECTRO/85. Conference Record. Los Angeles: Electronic Conventions Management; 1985; 13/4: 1-6.

Pilgrim, M. Millimetre-waves applied to TV distribution. In: IEE Colloquium on 'MM-wave and IR Applications, Devices and Propagation'. London: IEE; 1987; c4; 65: 1-4.

Poetsch, D. System approaches to HDTV telecine. In: IBC 86. International Broadcasting Convention. London: IEE; 1986: 229-31.

Powers, K.H. Enhancing quality in television systems-a glimpse of the future. In: IREECON International Sydney 83. 19th International Electronics Convention and Exhibition. Digest of Papers. Sydney, Australia: Institute of Radio and Electronic Engineers of Australia; 1983: 61-2.

Powers, K.H. Status and issues in HDTV production standards. In: Third International Colloquium on Advanced Television Systems: HDTV '87. Colloquium Proceedings. Montreal, Quebec, Canada: CBC; 1988; 1: 2/1/1-7.

Prodan, R.S. Multidimensional digital signal processing for high definition television. In: Third International Colloquium on Advanced Television Systems: HDTV '87. Colloquium Proceedings. Montreal, Quebec, Canada: CBC; 1988; 1: 6/4/1- 28.

Prodan, R.S. Multidimensional digital signal processing for high definition television. In: IEEE 1988 International Conference on Consumer Electronics. Digest of Technical Papers. New York: IEEE; 1988: 14.

Protter, H.E. Terrestrial broadcasting perspective on HDTV delivery. In: IEEE EASCON '88. 21st Annual Electronics and Aerospace Conference: How will Space and Terrestrial Systems Share the Future? Conference Proceedings. New York: IEEE; 1988: 179.

Reitmeier, G.A. Digital television-what does the future have in store? In: ELECTRO/85. Conference Record. Los Angeles: Electronic Convention Management; 1985; 13/0: 1-4.

Reuter, T. High definition television standards conversion. In: Second International Conference on Image Processing and its Applications. London: IEE; 1986: 224-8.

Reuter, T. Motion adaptive downsampling of high definition television signals. In: Proceedings of SPIE - The International Society for Optical Engineering. Bellingham, WA: SPIE; 1986; 594: 30-40.

Reuter, T. Multi-dimensional adaptive sampling rate conversion. In: Proceedings - ICASSP 87, 1987 IEEE International Conference on Acousitics, Speech and Signal Processing. New York: IEEE; 1987: 1969-72.

Rhodes, C.W. Time division multiplex of time compressed chrominance for a compatible high definition television system. In: IEEE Transactions on Consumer Electronics. New York: IEEE; 1982.

Ricaud, J.L. PRIMICON pick-up tube for EUREKA HDTV broad casting program. In: IBC 1988: International Broadcasting Convention. London: IEE; 1988: 181-4.

Richter, H.-P. Interfacing and signal distribution in component studios. In: IBC 1988: International Broadcasting Convention. London: IEE; 1988: 86-9.

Rilly, G. et al. Power concept for HDTV. In: IEEE 1988 International Conference on Consumer Electronics. Digest of Technical Papers. New York: IEEE; 1988: 196-7.

Robert, P.; Lamnabhi, M.; Lhuillier, J.J. Advanced high definition 50 to 60 Hz standard conversion. In: IBC 1988: International Broadcasting Convention. London: IEE; 1988: 21-4.

Roscoe, O.S. State of development of direct broadcasting satellite television. In: Canadian Satellite User Conference, 1987 - Conference Proceedings. Ottawa, Canada:

Telesat; 1987: 43-47.

Rossi, J.P.; McMann, R.H.; Goldberg, A.A. A two channel compatible HDTV broadcast system. In: DeWilde, P.; May, C.A. Links for the Future. Science, Systems and Services for Communications. Proceedings of the International Conference on Communications-ICC 84. Amsterdam, Netherlands: North-Holland; 1984; 3: 1092-3.

Rossi, J.P.; Goldberg, A.A.; McMann, R.H. Two channel compatible HDTV broadcast system. In: ICC '84 Conference Record. Amsterdam: Elsevier Science Publishers; 1984.

Sabatier, J. HDTV: a safe path? In: IBC 86. International Broadcasting Convention. London: IEE; 1986: 5-7.

Sablatash, M. Applications and future for digital processing in improved, extended and high definition TV systems. In: Third, International Colloquium on Advanced Television Systems:HDTV '87. Colloquium Proceedings. Montreal, Quebec, Canada: CBC; 1988; 1: 6/1/1-30.

Salvadorini, R. The problem of television standards. In: Proceedings of the 32nd Congress on Electronics: Satellite Broadcasting. Rome, Italy: Rassegna Int. Elettronica Nucl. & Aerospaziale; 1985: 231-42.

Sandbank, C.P.; Stone, M.A. The role of DATV in future television emission and reception. In: Third International Colloquium on Advanced Television Systems: HDTV '87. Colloquium Proceedings. Montreal, Quebec, Canada: CBC; 1988; 1: 4/4/1- 15.

Sawano, T. et al. Photonic switching system for broadband services. In: GLOBECOM Tokyo '87. IEEE/IECE Global Telecommunications Conference 1987. Conference Record. New York: IEEE; 1987; 3: 2039-43.

Schafer, R.; Kauff, P.; Chen, S.C. Technical solutions for a sequential HDTV-production standard with today's technology. In: Third International Colloquium on Advanced Television Systems: HDTV '87. Colloquium Proceedings. Montreal, Quebec, Canada: CBC; 1988; 1: 2/5/1-20.

Schamel, G. Multi dimensional interpolation of progressive frames from spatio-temporally subsampled HDTV fields. In: Proceedings - ICASSP 87, 1987 International Conference on Acoustics, Speech, and Signal Processing. New York: IEEE;

1987: 1965-68.

Schiffler, W.; Fehlauer, E. An HDTV VTR: analog recording with digital signal processing. In: IBC 1988. International Broadcasting Convention. London: IEE; 1988: 414-17.

Schonfelder, H. From component studio to HDTV production. In: TELEMATICA 88: Internationaler Fachkongress fur Integrierte Telekommunikation, Telematik, Kabel-und Satellitenkommunikation (TELEMATICA 88: International Conference for Integrated Telecommunications, Telematics, Cable and Satellite Communications). Munich, West Germany: Verlag Reinhard Fischer; 1988: 567-77.

Schreiber, W.F. 6-MHz single-channel HDTV systems. In: Third International Colloquium on Advanced Television Systems: HDTV '87. Colloquium Proceedings. Montreal, Quebec, Canada: CBC; 1988; 2: 4/7/1-12.

Searle, R.P. et al. M**3VDS - millimetre-wave TV distribution - the system and its technology. In: Second IEE National Conference on Telecommunications. London: IEE; 1989; c300: 202-207.

Segar, P.L. The DMAC packet transmission system for UK DBS. In: IEE Colloquium on 'The UK Direct Broadcast Satellite'. London: IEE; 1989: 5/1-7.

Sewter, J.B. Communication and direct broadcasting satellites. In: International Conference on the History of Television - From Early Days to the Present. London: IEE; 1986: 181-5.

Shibaya, H. et al. The wide band ferrite head: its application to the high definition video recording. In: Watanabe, H.; Iida, S.; Sugimoto, M. Ferrites. Proceedings of the ICF 3. Third International Conference on Ferrites. Dordrecht, Netherlands: Reidel; 1982: 685-9.

Silzars, A.K. Display marketplace in 1996. In: Society for Information Display - 1986 Seminar Lecture Notes. Playa del Rey, CA: Society for Information Display; 1986: 5. 1. 1-5. 1. 25.

Smith, J.M.; Stewart, J. From vacuum tubes to solid state imagers - the past, the present, the future. In: International Conference on the History of Television - From Early Days to the Present. London: IEE; 1986; c271: 73-76.

Snelling, R.K. Fiber optic industry perspective on HDTV delivery.

In: IEEE EASCON '88. 21st Annual Electronics and Aerospace Conference: How will Space and Terrestrial Systems Share the Future? Conference Proceedings. New York: IEEE; 1988: 187-8.

Stallard, W.G. Compatible extended definition TV, an available approach. In: 1984 IEEE international conference on consumer electronics. Digest of technical papers. New York: IEEE; 1984: 120-1.

Stechele, W.; Ruge, I. Microelectronics and its influence on telecommunication. In: 7th European Conference on Electrotechnics: Advanced Technologies and Processes in Communication and Power Systems. EUROCON 86. Paris: Comite EUROCON 86; 1986 April: 330-2.

Storey, R. HDTV motion adaptive bandwidth reduction using DATV. In: IBC 86. International Broadcasting Convention. London: IEE; 1986: 167-72.

Storey, R. Motion compensated DATV bandwidth compression for HDTV. In: IBC 1988: International Broadcasting Convention. London: IEE; 1988: 78-81.

Storey, R. Signal processing and bandwidth compression techniques. In: IEE Colloquium on 'Progress Towards HDTV'. London: IEE; 1989: 2/1-4.

Sturt, A.H.J. A high definition television standard based on 72 frames per second. In: IBC 86. International Broadcasting Convention. London: IEE; 1986: 163-6.

Sugawara, M.; Kurai, T.; Uchiike, H. Surface-discharge color plasma display with common electrode structure. In: Morreale, J. 1988 SID International Symposium. Digest of Technical Papers. First Edition. Playa del Rey, CA: SID; 1988: 150-2.

Sugimoto, M. The NHK strategy for HDTV services. In: Third International Colloquium on Advanced Television Systems: HDTV '87. Colloquium Proceedings. Montreal, Quebec, Canada: CBC; 1988; 1: 5/1/1-17.

Sun, Y.; Gerla, M. SFPS: a synchronous fast packet switching architecture for very high speeds. In: Proceedings - IEEE INFOCOM. Piscataway, NJ: IEEE; 1989; II: 641-646.

Switzer, I. A 'perfect picture' service for cable. In: CATCOM. Erster internationaler kabelfernseh-kongress (First international cable television congress. Bern, Switzerland:

Verband Schweizerischer Kabelfernsehbetriebe; 1985:
B19/1- 7.

Tadokoro, Y. CCIR activities for standardization of high
definition television system. In: IBC 86. International
Broadcasting Convention. London: IEE; 1986: 8-11.

Tanaka, Y.; Kubota, K.; Iwadate, Y. Transmission of HDTV signals
by direct broadcasting satellite and communication
satellite. In: IBC 1988: International Broadcasting
Convention. London: IEE; 1988: 41-4.

Tanimoto, M. et al. A new bandwidth compression system of picture
signals-the TAT. In: GLOBECOM '85. IEEE Global Telecommunications Conference. Conference Record. Communication Technology to Provide New Services. New York:
IEEE; 1985; 1: 427-31.

Tanimoto, M. et al. The TAT system for high quality image compression. In: GLOBECOM Tokyo '87. IEEE/IECE Global Telecommunications Conference 1987. Conference Record. New
York: IEEE; 1987; 1: 416-20.

Tanimura, H.; Hashimoto, Y.; Yoshinaka, T. HDTV digital tape
recording. In: Third International Colloquium on Advanced
Television Systems: HDTV '87. Colloquium Proceedings.
Montreal, Quebec, Canada: CBC; 1988; 1: 2/7/1-19.

Tanton, N.E. HDTV sources for the evaluation of studio and transmission standards. In: IBC 86. International Broadcasting
Convention. London: IEE; 1986: 353-7.

Tanton, N.E.; Stone, M.A. HDTV displays: subjective effects of
scanning standards and domestic picture sizes. In: IBC 1988:
International Broadcasting Convention. London: IEE; 1988:
204-11.

Thomas, W.; Heimbach, P. Cable's role in the development of HDTV
systems. In: IEEE 1988 International Conference on Consumer Electronics. Digest of Technical Papers. New York:
IEEE; 1988: 258-9.

Thompson, M. Psychophysics and future television technology. In:
Electronic Imaging '88: International Electronic Imaging
Exposition and Conference. Advance Printing of Paper
Summaries. Waltham, MA: Institute of Graphic Communica
tion; 1988; 1: 106-8.

Thorpe, L.; Ozaki, Y. HDTV electron beam recording. In: Third

International Colloquium on Advanced Television Systems: HDTV '87. Colloquium Proceedings. Montreal, Quebec, Canada: CBC; 1988; 1: 2/8/1-17.

Todorovic, A. HDTV production standard-the European scene. In: Third International Colloquium on Advanced Television Systems: HDTV '87. Colloquium Proceedings. Montreal, Quebec, Canada: CBC; 1988; 1: 2/2/1-13.

Tolkoff, J. High def in Paris. In: Electro/87 and Mini/Micro Northeast Conference Record. Los Angeles: Electronic Conventions; 1987: 33.2. 1-33.2.3.

Tomoda, K. et al. Automatic convergence alignment system for 400-inch rear projection HDTV. In: 1986 IEEE International Conference on Consumer Electronics. Digest of Technical Papers. ICCE. New York: IEEE; 1986: 244-5.

Tonge, G.; Buckley, J.; Long, T.J. HDTV 1986-the year of decision in perspective. In: IBC 86. International Broadcasting Convention. London: IEE; 1986: 144-8.

Tonge, G.J.; Childs, I. The relationship between high definition television and DBS video transmission. In: Colloquium on 'Better Television by Satellite - Receiver and Modulation Techniques'. London: IEEE; 1983; 1-4.

Tonge, G.J. Extended definition television through digital signal processing. In: IBC 82. International Broadcasting Convention. London: IEE; 1982: 148-51.

Tonge, G.J. Prospects of higher quality pictures via direct broadcast satellites. In: Cantraine, G.; Destine, J. New Systems and Services in Telecommunications II. Proceedings of the Second International Conference. Amsterdam, Netherlands: North-Holland; 1984: 393-9.

Tonge, G.J. Time-sampled motion portrayal. In: Second International Conference on Image Processing and its Applications. London: IEE; 1986: 215-19.

Tonge, G.J. Wide aspect ratio MAC. In: Cantraine, G.; Destine, J. New Systems and Services in Telecommunications, III: Networks, Cables, Satellites - The What, the How, the Why? Proceedings of the Third International Conference. Amsterdam: North-Holland; 1987: 313-20.

Toth, A.G. Hierarchical evolution of high definition television. In: GLOBECOM Tokyo '87. IEEE/IECE Global Telecommunications Conference 1987. Conference Record. New York: IEEE; 1987; 1: 405-10.

Toth, A.G. Hierarchical evolution of high defintion television. In: Third International Colloquium on Advanced Television Systems: HDTV '87. Colloquium Proceedings. Montreal, Quebec, Canada: CBC; 1988; 1: 5/3/1-15.

Toth, A.G.; Tsinberg, M.; Rhodes, C.W. Hierarchical NTSC compatible HDTV system architecture-a North American perspective. In: IBC 1988: International Broadcasting Convention. London: IEE; 1988: 30-3.

Toth, A.G. High definition television-a North American perspective. In: FOC/LAN '88 Proceedings. The Twelfth International Fiber Optic Communications and Local Area Networks Exposition. Boston: Gatekeepers; 1988: 331-7.

Treurniet, W.C. Estimating movement direction with a neural network. In: Proceedings of Vision Interface '88. Toronto, Ontario, Canada: Canadian Image Processing & Pattern Recognition Society; 1988: 110-14.

Trew, T.I.P.; Moris, O.J. Spatially adaptive sub-branches for HD-MAC. in: IBC 1988: International Broadcasting Convention. London: IEE; 1988: 66-9.

Tsinberg, M. NTSC compatible HDTV emission system. In: Third International Colloquium on Advanced Television Systems: HDTV '87. Colloquium Proceedings. Montreal, Quebec, Canada: CBC; 1988; 1: 4/6/1-16.

Turk, W.E. Television cameras-the sensor choice. In: IBC 82. International Broadcasting Convention. London: IEE; 1982: 24-8.

Tzou, K.-H et al. Compatible HDTV coding for broadband ISDN. In: GLOBECOM '88. IEEE Global Telecommunications Conference & Exhibition: Communications for the Information Age - Conference Record. New York: IEEE; 1988: 743-49.

Uchiike, H. et al. Improved electrode structure and driving method for color surface-discharge AC plasma display panels. In: 1986 SID International Symposium. Digest of Technical Papers. First Edition. New York: Palisades Institute of Research Services; 1986: 216-19.

Umemoto, M. et al. High data rate recording for HDTV digital VTR. In: The 4th Joint MMM-Intermag conference. Piscataway, NJ:

IEEE; 1988.

Umemoto, M.; Mita, S.; Eto, Y. High data rate recording for a high definition digital VTR. In: Video, Audio and Data Recording. Sixth International Conference. London: IERE; 1986: 81-85.

Vivian, W.E. Financial implications of alternate scenarios for worldwide entry of high-definition television services. In: Third International Colloquium on Advanced Television Systems: HDTV '87. Colloquium Proceedings. Montreal, Quebec, Canada: CBC; 1988; 1: 5/7/1-16.

Vorstermans, J.; Theeuws, R. Service analysis for the customers premises network. In: ISSLS 88: The International Symposium on Subscriber Loops and Services. Proceedings. New York: IEEE; 1988: 103-7.

Vreeswijk, F.W.P. et al. HD-MAC coding of high definition television signals. In: IBC 1988: International Broadcasting Convention. London: IEE; 1988: 62-5.

Wassiczek, N. HDTV-report on the situation in Europe. In: Third International Colloquium on Advanced Television Systems: HDTV '87. Colloquium Proceedings. Montreal, Quebec, Canada: CBC; 1988; 1: 5/5/1-5.

Weese, D.E.; Gray, D. Transmission of HDTV demonstrations during HDTV '87. In: Third International Colloquium on Advanced Television systems: HDTV '87. Colloquium Proceedings. Montreal, Quebec, Canada: CBC; 1988; 1: 3/3/2-10.

Weiss, S.M. Migrating to advanced television in the United States. In: IBC 1988: International Broadcasting Convention. London: IEE; 1988: 425-8.

Weissensteiner, W. Concept of a consumer-type HDMAC VCR. In: IBC 1988: International Broadcasting Convention. London: IEE; 1988: 228-30.

Wendland, B. A hierarchical TV system for different communication services. In: DeWilde, P.; May, C.A. Links for the Future. Science, Systems and Services for Communications. Proceedings of the International Conference on Communications. Amsterdam, Netherlands: North-Holland; 1984; 1: 724-7.

Wendland, B. High quality television by signal processing. In: Cantraine, G.; Destine, J. New Systems and Services in

Telecommunications II. Proceedings of the Second
International Conference. Amsterdam: North-Holland; 1984:
401-9.

Wendland, B. On picture scanning for future HDTV-systems. In: IBC
82. International Broadcasting Convention. London: IEE;
1982: 144-7.

Westerkamp, D.; Keesen, H.-W.; Plantholt, M. The benefits of a
progressive scan based HDTV system. In: IEEE 1988
International Conference on Consumer Electronics. Digest of
Technical Papers. New York: IEEE; 1988: 190-1.

Weston, M.; Ackroyd, D.M. Fixed, adaptive, and motion compensated interpolation of interlaced TV pictures. In: IBC 1988:
International Broadcasting Convention. London: IEE; 1988:
220-3.

Wilcock, P.E. HDTV: how much is enough? In: IEE Colloquium on
'HDTV Bandwidth Reduction'. London: IEE; 1987; c4;
24: 1-8.

Willard, R.A. Television sound equipment. In: International Conference on the History of Television - From Early Days to the
Present. London: IEE; 1986: 158-60.

Windram, M.D.; Morcom, R.; Wilson, E.J. Ensuring an evolution to
high definition television for satellite broadcasting. In:
IBC 86. International Broadcasting Convention. London: IEE;
1986: 32-6.

Windram, M.D.; Drury, G.M. Toward high definition television. In:
IBC 1988: International Broadcasting convention. London:
IEE; 1988: 1-7.

Windram, M.D.; Tonge, G.J. The D-MAC transmission system for
enhanced and high definition television services. In: Third
International Colloquium on Advanced Television Systems:
HDTV '87. Colloquium Proceedings. Montreal, Quebec,
Canada: CBC; 1988; 1: 4/2/1-13.

Wood, D. Subjective quality assessment today-new routes form old
roots? (HDTV). In: Third International Colloquium on
Advanced Television Systems: HDTV '87. Colloquium
Proceedings. Montreal, Quebec, Canada: CBC; 1988; 1: 1/1/
1- 11.

Wood, D.; Habermann, W. The EBU contribution to the world-wide
HDTV study. In: IBC 86. International Broadcasting

Convention. London: IEE; 1986: 149-52.

Wood, D.; Jones, B.L. The evaluation of narrow-band HDTV transmission standards such as HD-MAC. In: IBC 1988: International Broadcasting Convention. London: IEE; 1988: 49-53.

Wood, D.; Jones, B.L. Subjective assessments-good luck or good planning? In: Television Measurements (Broadcasting and Distribution). Third International Conference. London: IERE; 1987: 3-7.

Yamazaki, S. et al. Tunable optical heterodyne receiver for coherent FDM broadcasting systems. In: Fourteenth European Conference on Optical Communication (ECOC 88). London: IEE; 1988; 1: 86-9.

Yashima, Y.; Sawada, K. A highly efficient coding method for HDTV signals. In: IEEE International Conference on Communications '87: Communications-Sound to Light. Proceedings. New York: IEEE; 1987; 1: 125-9.

Yoshimoto, S. et al. A trade-off study on 22 GHz-band multibeam satellite broadcasting systems. In: AIAA 11th Communication Satellite Systems Conference. Collection of Technical Papers. New York: AIAA; 1986: 225-36.

Yoshimura, H.; Nakanishi, T.; Tamauchi, H. A 50 MHz CMOS geometrical mapping processor. In: Winner, L. 1988 IEEE International Solid-State Circuits Conference. Digest of Technical Papers. 31st ISSCC First Edition. Coral Gables, FL: Lewis Winner; 1988 February: 162-3, 347-8.

Ziemer, A. HDTV, from studio to receiver-a systematic analytical view. In: TELEMATICA 88: Internationaler Fachkongress fur Integrierte Telekommunikation, Telematik, Kabel-und Satellitenkommunikation (TELEMATICA 88: International Conference for Integrated Telecommunications, Telematics, Cable and Satellite Communications). Munich, West Germany: Verlag Reinhard Fischer; 1988: 578-87.

Zimmermann, R. Hardware and software viewdata. In: Bildschirmtext Kongress 1980 (Teletext Congress 1980). Frankfurt, Germany: Diebold Deutschland GMBH; 1980: 117-38.

Part Three: Newspapers and News Magazines

Abramowitz, M. Broadcasters woo with crisper shots. In: Washington Post. Washington, DC; 1987 January 4; sec. H: 1.

Allison, D. Turnaround rocking Sci-Atlanta. In: Atlanta Business Chronicle. Atlanta; 1989 August 21: 1A.

Alster, N. TV's high stakes, high-tech battle. Fortune; 1988 October 24: 161-63, 166, 170.

Alster, N.; Katayama, F.H. TV's high-stakes, high-tech battle. Fortune; 1988 October 24; 118: 161-65.

Angus, R. HDTV: the FCC edit. High Fidelity; 1989 January; 39: 12-13.

Anonymous. American survey: suffering from decline? Try the consortium cure. Economist (Great Britain); 1989 March 25; 310(7595): 25-6.

Anonymous. America's billion-dollar boob-tube battle. The Economist (Great Britain); 1989 May 27; 311: 67-68.

Anonymous. Applications for the grants. In: New York Times. New York; 1989 March 9; sec. B: 23.

Anonymous. B'casters warned to move fast for hi-def TV benefits. Variety; 1988 January 13; 329: 1-2.

Anonymous. The big picture. Time; 1989 March 13; 133: 49.

Anonymous. Broadcasters support HDTV blueprint. Variety; 1988 December 7; 333: 3.

Anonymous. But: waiving antitrust rules means new rules. In: New York Times. New York; 1989 May 15; sec. A: 18.

Anonymous. Cable consortium formed. In: New York Times. New York; 1988 May 6; sec. D: 20.

Anonymous. Cablers envision multi-standard HDTV sets. High

Technology Business; 1988 December; 8: 38-39.

Anonymous. CBS-TV takes high-definition test to H'wood. Variety; 1982 January 27; 305: 47.

Anonymous. The CEOs of Minnesota. In: Corporate Report Minnesota. Minneapolis, MN; 1989 June 21: 35.

Anonymous. CoMED HMO: marking its 10th anniversary. In: Business Journal of New Jersey. Morganville, NJ; 1988 February: 32.

Anonymous. Coming: picture perfect TV. Changing Times; 1984 May; 38: 73-74.

Anonymous. Consumers will go for HDTV says EIA study. Broadcasting; 1988 December 5; 115: 64.

Anonymous. DARPA to fund high-density TV. High Technology Business; 1989 April; 9: 33.

Anonymous. Do not adjust your set. Economist (Great Britain); 1988 October 1; 309: 80-1.

Anonymous. Europe pushes sharper TV look. In: New York Times. New York; 1988 November 15; sec. D: 22.

Anonymous. Europe seeks TV backing. In: New York Times. New York; 1989 February 22; sec. D: 22.

Anonymous. Firm joins group exploring advanced-TV technology. In: Wall Street Journal. New York; 1989 January 18; sec. B: 4.

Anonymous. Go tell Ma Bell 'I want my HDTV'. PC Week; 1988 August 15; 5: C22.

Anonymous. HD no affair of broadcasters, asserts topper of 1125 productions. Variety; 1988 October 12; 332: 74.

Anonymous. HDTV. Video Review; 1989 January; 9: 53-57.

Anonymous. HDTV - an alternative viewpoint. Radio-Electronics; 1989 February; 60: 78-83.

Anonymous. HDTV as security, economic issue: government support not coming. Television/Radio Age; 1989 March 20; 36: 59-61.

Anonymous. HDTV backing for five firms. In: Computergram International; 1989 June 15; (1199): CGIo6150002.

Anonymous. HDTV is taking off in Japan: theaters using it on launch pad. Variety; 1988 February 17; 330: 161.

Anonymous. HDTV plea on Soviets. In: New York Times. New York; 1989 March 21; sec. D: 10.

Anonymous. HDTV transmission standard delayed. Broadcasting;

1989 April 24; 116: 119-120.

Anonymous. HDTV: a better buggy whip. National Review; 1989 July 14; 41: 14-15.

Anonymous. HDTV: will U.S. be in the picture? In: New York Times. New York; 1988 September 21; sec. D: 1.

Anonymous. HDTV: fund research, not production. Business Week; 1989 May 22: 188.

Anonymous. High definition TV myths debunked. High Technology Business; 1988 August: 55-57.

Anonymous. High definition's high visibility in Las Vegas. Broadcasting; 1989 May 8: 32-33.

Anonymous. High definition TV a focus of tension. In: San Francisco Chronicle. San Francisco; 1987 December 26; sec. B: 3.

Anonymous. High-definition TV: breakthrough, boondoggle--or both. Consumer Reports; 1989 October; 54(10): 627-29.

Anonymous. High-definition TV venture. In: New York Times. New York; 1989 March 15; sec. B: 5.

Anonymous. High-definition mercantilism. In: Christian Science Monitor. Boston; 1989 May 18: 20.

Anonymous. High-definition TV hearings. In: New York Times. New York; 1989 March 23; sec. D: 8.

Anonymous. High-definition, high-stakes TV. In: Boston Globe. Boston; 1988 December 28: 10.

Anonymous. High-tech TV standard is affirmed. In: New York Times. New York; 1989 February 3; sec. D: 15.

Anonymous. How soon the supertelly? The Economist (Great Britain); 1988 January 30; 306: 70-1.

Anonymous. How to look at TV, and the future. In: New York Times. New York; 1988 November 1; sec. A: 30.

Anonymous. Japan pubcaster to pitch HDTV in D.C.: quiet fears about trade. Variety; 1988 August 17; 332: 2.

Anonymous. Japan prepares for EDTV broadcasting. Television Digest; 1988 October 17; 28: 12-15.

Anonymous. Japan plunges ahead on high definition TV. In: Chicago Tribune. Chicago; 1987 December 26; sec. 2: 8.

Anonymous. Kitchen's recipe has TTC cooking again. In: Denver Business. Denver, CO; 1989 April: 16.

Anonymous. Low on high resolution. In: Chicago Tribune. Chicago;

1988 May 9; sec. 4: 1.

Anonymous. Low-cost HDTV that beats Japan's? Business Week; 1989 May 29: 102.

Anonymous. NASA precedes Pentagon into HDTV research. Broadcasting; 1989 January 2; 116: 94-5.

Anonymous. NBC unveils system for high-quality TV. In: San Francisco Chronicle. San Francisco; 1987 October 6; sec. C: 4.

Anonymous. Networks oppose an HDTV standard not compatible with existing TV sets. In: Atlanta Constitution. Atlanta, GA; 1988 June 24; sec. B: 6.

Anonymous. New Media. Japan Marketing/Advertising Yearbook (Japan); 1987: 147-52.

Anonymous. New TV may bring new habits in viewing. In: Washington Times. Washington, DC; 1987 November 19; sec. C: 4.

Anonymous. NTIA director sez HDTV needs govt.-industry support to work. Variety; 1987 October 7; 328: 82.

Anonymous. Pentagon help in TV research. In: New York Times. New York; 1988 December 19; sec. D: 5.

Anonymous. Pentagon aims to give U.S. TV makers a boost. In: Chicago Tribune. Chicago; 1988 December 20; sec. 1: 10.

Anonymous. Pitch for technology aid. In: Chicago Tribune. Chicago; 1988 September 8; sec. 3: 1.

Anonymous. Prospects for the new generation of television systems: high defintion TV and extended-definition TV. Japanese Finance and Industry (Japan); 1989 June: 1-15.

Anonymous. The race for HDTV. Design News; 1989 July 3; 45: 202.

Anonymous. Raychem, Xerox plan venture to develop large HDTV screen. In: Wall Street Journal. New York; 1989 March 15; sec. B: 5.

Anonymous. Report says U.S. activity vital in high-definition TV market. In: Atlanta Constitution. Atlanta, GA; 1988 November 25; sec. B: 6.

Anonymous. Research spending in U.S. set on high-definition TV. In: Wall Street Journal. New York; 1989 April 28; sec. B: 2.

Anonymous. See mixed market IC market taking off. Electronic News; 1989 July 10; 35(1766): 32.

Anonymous. Shimmer, flicker and picture tricks. The Economist (Great Britain); 1988 January 30; 306: 71-72.

Anonymous. Should Uncle Sam tilt to HDTV? Yes: it's pivotal to industries of the future. In: New York Times. New York; 1989 May 15; sec. A: 18.

Anonymous. SIGGRAPH '89 preview. In: Computer Graphics World; 1989 July; (7): 166-77.

Anonymous. Sony KX-1901A video monitor. High Fidelity; 1984 January; 34: 47-49.

Anonymous. Sony to expand research center. In: New York Times. New York; 1989 April 28; sec. D: 5.

Anonymous. South Korea details plans for high-definition TV. In: Wall Street Journal. New York; 1989 June 27; sec. A: 13.

Anonymous. SRI lands California grant to develop cold cathode flat panel display technology. In: Computergram International; 1989 July 24; (1226): CGI07240008.

Anonymous. Super television: the high promise--and high risks-- of high-definition TV. Business Week; 1989 January 30: 56- 9+.

Anonymous. Taking the long view on high-definition TV. Business Week; 1989 January 30: 104.

Anonymous. Television and the economy: an interview with William F. Schreiber. Technology Review; 1989 April; 92: 35-37.

Anonymous. Television on hold. The Economist (Great Britain); 1988 October 1; 309: 17-18.

Anonymous. Toward a global high-definition TV production standard. Department of State Bulletin; 1989 June: 48-51.

Anonymous. Trumbull taking high-D road on innovative 'Roses' short: forsees blend of TV and film. Variety; 1989 June 14; 335: 35.

Anonymous. TV format questioned. In: New York Times. New York; 1989 August 3; sec. D: 14.

Anonymous. U.S.-Europe TV proposal. In: New York Times. New York; 1989 May 17; sec. C: 4.

Anonymous. Ultravideo. High Fidelty; 1989 April: 47-48, 50, 56.

Anonymous. Vacuum electronics could find use in HDTV technology. Research & Development; 1989 August; 31: 33-34.

Anonymous. Videotape system may improve the picture quality of television. In: New York Times. New York; 1988 April 13; sec. D: 6.

Anonymous. White House to recommend easing laws on antitrust to spur high-d development. Variety; 1989 May 10; 335(94).

Anonymous. Zenith should stay tuned in to TVs. Business Week; 1988 October 3: 146.

Anonymous. Zenith Electronics chief seeks U.S. tax for HDTV research. In: Wall Street Journal. New York; 1989 May 17; sec. B: 3.

Anonymous. Zenith topper urges tax on new TV sets to fund HD research. Variety; 1989 May 24; 335: 57.

Areddy, J.T. U.S. HDTV project gets bids of units of foreign firms. In: Wall Street Journal. New York; 1989 March 6; sec. B: 2.

Auerbach, S. High-definition Commerce chief. In: Washington Post. Washington, DC; 1989 April 28; sec. A: 23.

Auerbach, S. U.S.-EC joint effort on HDTV to be urged: W. German to propose combining forces to overcome Japan's lead. In: Washington Post. Washington, DC; 1989 May 16; sec. B: 3.

Barfield, C. It's still high-definition intervention. In: Wall Street Journal. New York; 1989 May 8; sec. A: 16.

Barrett, J. Debate role of government in consortiums. Electronic News; 1989 June 12; 35(1762): 13.

Barrett, J. MIT accused of selling U.S.-funded data. Electronic News; 1989 June 19; 35(1763): 12.

Behrens, S. The fight for high-def. Channels of Communications;1986 May; 6: 42-46.

Behrens, S. High-definition's spectrum needs spur TV broadcasters to action. Channels The Business of Communications; 1987 April; 7: 16.

Behrens, S. Passage to high-def. Channels The Business of Communications; 1988 April; 8: 54-56.

Behrens, S. Politics muddies the clear picture: tomorrow's TV may still need conversion to cross borders. Channels The Business of Communications; 1986 December; 6: 54.

Behrens, S. Suddenly, TV's overhaul gets priority treatment: broadcast and cable leaders are looking closely at the picture of the future - and even more closely at each other's moves. Channels The Business of Communication;1987 December; 7: 90.

Belsky, G. Electronics stores are plugging in again. In: Crains New York Business. New York; 1989 January 16: 3.

Benzon, W. New generation of TV brings along new host of challeng-

es. In: Capital Business Review. Albany, NY; 1988 February
1: 17.

Berger, M. Japan expects HDTV boom. In: San Francisco Chronicle.
San Francisco; 1988 May 31; sec. C: 1.

Bernard, J. High definition TV. Radio-Electronics; 1987 August;
58: 48-51.

Bettelheim, A. U.S. chases Europe, Japan in 'super-TV' technology.
In: Denver Post. Denver, CO; 1989 April 3; sec. D: 1.

Blinder, E. Closing in on HDTV: the FCC took a major step toward
an advanced TV standard. Why is there still such confusion?
Channels The Business of Communications; 1988 December;
8: 54-55.

Blinder, E.J. Innovations here and still to come. Channels the
Business of Communications; 1989 May; 9(46-48).

Booth, S.A. Catch a wave. Popular Mechanics; 1988 August; 165:
34-36.

Booth, S.A. Futurevision. Popular Mechanics; 1986 July; 163: 67- 69.

Brennan, P.L. Truevision's hardware offers high-resolution
graphics in 16 million colors. In: Indianapolis Business
Journal. Indianapolis, IN; 1988 March 7; sec. B: 1-3.

Brewin, B. HDTV, American style. Premiere; 1989 August; 2: 97.

Broad, W.J. U.S. counts on computer edge in the race for advanced
TV. In: New York Times. New York; 1989 November 28;
sec. C: 1.

Broder, J.M. Advanced technology held vital to American interests.
In: Los Angeles Times. Los Angeles; 1989 March 9; sec.
IV: 1.

Brody, H. TV - the push for a sharper picture. High Technology; 1988
April; 8: 25-29.

Brown, G.E., Jr. U.S. must get in race on high-definition TV. In: Los
Angeles Times. Los Angeles; 1988 October 14; sec. II: 7.

Bulkley, K. ATC will invest millions in fiber-optic network. In:
Denver Business Journal. Denver, CO; 1987 October 26: 11.

Bulkley, K. Cable firms pour millions into fiber optics. In:
Denver Business Journal. Denver, CO; 1989 February 13: 13.

Bulkley, K. Cable TV poised for growth in programming and
technology. In: Denver Business Journal. Denver, CO; 1988
January 4: 10.

Bulkley, K. Fiber-optic pioneers wage epic legal battle. In: Denver

Business Journal. Denver, CO; 1989 July 24: 1.

Bumgarner, T. Zenith amends bylaws to satisfy shareholders. In: Springfield Business Journal. Springfield, MO; 1988 October 3: 1.

Burgess, J. $1.35 billion sought for HDTV consortium; firms want government-industry cooperation on U.S. development. In: Washington Post. Washington, DC; 1989 May 10; sec. E: 1.

Burgess, J. Chances are high TV breakthrough is in the air; U.S. industry looks for its place in the picture even though Japanese have head start. In: Washington Post. Washington, DC; 1987 October 11; sec. K: 1.

Burgess, J. Commerce to drop role in HDTV. In: Washington Post. Washington, DC; 1989 September 13; sec. C: 1.

Burgess, J. Digital radio: will the public pay to listen? In: Washington Post. Washington, DC; 1988 July 28; sec. E: 3.

Burgess, J. The global race is on for next-generation TV: technical questions abound on 'high-definition'. In: Washington Post. Washington, DC; 1988 September 11; sec. H: 4.

Burgess, J. More funds are sought for HDTV; 87 proposals submitted for research backing. In: Washington Post. Washington, DC; 1989 March 9; sec. E: 1.

Burgess, J. Sony announces U.S. center for research into HDTV. In: Washington Post. Washington, DC; 1989 April 28; sec. F: 3.

Burgess, J. Sony requests U.S. funds for HDTV project. In: Washington Post. Washington, DC; 1989 March 3; sec. G: 3.

Burgess, J. Texas Instruments gets license for Japanese HDTV technology. In: Washington Post. Washington, DC; 1989 September 14; sec. E: 3.

Burgess, J. U.S. withdraws support for studio HDTV standards: Japanese suffer setback in global effort. In: Washington Post. Washington, DC; 1989 May 6; sec. D: 12.

Burgess, J. U.S. won't back Japan on high-definition TV. In: Los Angeles Times. Los Angeles; 1989 May 8; sec. IV: 4.

Burgess, J. Zenith, AT&T seek U.S. funds for joint venture on HDTV. In: Washington Post. Washington, DC; 1989 March 1; sec. E: 3

Burgess, J.; Richards, E. Can U.S. protect lead in supercomputers? In: Washington Post. Washington, DC; 1989 May 7; sec. H: 1.

Carlson, A. Making its mark: Indy's reputation in fiber optics continue to grow. In: Indianapolis Business Journal. Indianapolis, IN;

1989 July 17; sec. B: 1.

Cauley, L. Advanced TV given the go-ahead. In: Washington Times. Washington, DC; 1988 September 2; sec. C: 1.

Cavanaugh, T. Debate goes on to open competition for cable TV. In: Capital District Business Review. Albany, NY; 1988 October 17: 18.

Churbuck, D.; Gilder, G. IBM-TV? Forbes; 1989 February 20; 143: 72-75.

Cohen, C. Digital processing hikes resolution to sharpen TV image. Electronics; 1983 September 8; 56: 77-78.

Cohen, C.L. High-definition TV signal fits on one channel with interpolation of missing pixels. ElectronicsWeek; 1984 February 23; 57: 74-75.

Cohen, C.L. NEC builds imager for HDTV: it wasn't easy. Electroics; 1986 March 3; 59: 16-17.

Cohen, C.L. NTT and NHK hone systems for sharp TV. Electronics; 1983 July 14; 56: 82-83.

Cohen, C.L. Upgraded analog monitor yields high-resolution picture. ElectronicsWeek; 1984 October 8; 57: 34-35.

Connelly, J. $1.75B super CPU network backed. Electronic News; 1989 July 31; 35(1769): 4.

Connelly, J. DARPA picks five firms for HDTV contracts. Electronic News; 1989 June 19; 35(1763): 6.

Connelly, J. HDTV--Who speaks for U.S. industry? Electronic News; 1989 January; 35(1-3).

Cook, W.J. Making a leap in TV technology. U.S. News & World Report; 1989 January 23; 106: 48-49.

Cooper, B., Jr. High-definition DBS. Radio-Electronics; 1987 July; 58: 62-63.

Cooper, B., Jr. Is HDTV the key to an international standard? Radio-Electronics; 1987 August; 58: 28-30.

Cooper, V. Digital graphics: wave of the future. In: Boulder County Business Report. Boulder, CO; 1987 August: 1.

Corcoran, E. Signing off? Tune in for the next episode in the television saga. Scientific American; 1988 October; 259: 138-9.

Costello, M. A poor man's high-def: U.S. broadcasters should take note of Japan's coming low-cost alternative to HDTV. Channels The Business of Communications; 1989 April;

9: 64.

Crane, R.J. Making America competitive: high definition TV. In: Chicago Tribune. Chicago; 1988 October 3; sec. 1: 13.

Crane, R.M. Advanced television: an American challenge. In: Boston Globe. Boston; 1988 November 8: 46.

Davis, B. AT&T manager on House panel writes TV bill. In: Wall Street Journal. New York; 1989 March 28; sec. B: 4.

Davis, B. FCC is freezing some TV requests, citing technology. In: Wall Street Journal. New York; 1987 July 17: 25.

Davis, B. FCC sets rules for broadcasting movie-quality TV. In: Wall Street Journal. New York; 1988 September 2: 34.

Davis, B. Firms plan to ask U.S. for millions to develop HDTV. In: Wall Street Journal. New York; 1989 May 5; sec. A: 16.

Davis, B. Funding for high-definition TV by the Pentagon may be boosted. In: Wall Street Journal. New York; 1989 March 9; sec. B: 4.

Davis, B. High definition TV contracts awarded by defense agency. In: Wall Street Journal. New York; 1989 June 14; sec. B: 4.

Davis, B. Pentagon seeks to spur U.S. effort to develop 'high-definition' TV. In: Wall Street Journal. New York; 1989 January 5; sec. B: 2.

Davis, B. U.S. is asked for subsidies of TV research; electronics group's request for $1.35 billion ignites industrial-policy battle. In: Wall Street Journal. New York; 1989 May 10; sec. A: 2.

Davis, B. Will high-definition TV be a turn-off? Perhaps, thanks to cost of sets and screen size. In: Wall Street Journal. New York; 1989 January 20; sec. B: 1.

Diebold, J.; Reich, R.B.; Phillips, K.I. How to get back into the TV technology race. In: New York Times. New York; 1988 November 11; sec. A: 30.

Donahue, H.C. Choosing the TV of the future: the stakes are high in the search for a broadcast standard for high-definition television. Technology Review; 1989 April; 92: 30-36.

Donlan, T.G. Redoubtable DARPA: it shapes the future of U.S. technology. Barron's; 1989 April 3; 69(14): 14-15, 18- 22.

Dorcoran, E. A technological fix: the U.S. searches for a stand on technology. Scientific American; 1989 August; 261: 60.

Dorland, M. The imperial image: notes on technology as ideology. Cinema Canada; 1985 January: 7-11.

Dryden S. Helping hand, not a handout: U.S. can spur industry's development of HDTV without giving away billions of tax-payer dollars. In: Los Angeles Times. Los Angeles; 1989 May 28; sec. IV: 3.

Dwyer, J., III. Southwestern Bell demonstrates high definition television. In: St. Louis Business Journal. St. Louis, MO; 1988 August 8; sec. A: 15.

Eckhouse, J. New wave of TVs to premiere. In: San Francisco Chronicle. San Francisco; 1988 May 31; sec. C: 1.

Falzone, K. Investors fume as Zenith vies for future TV niche. In: Crains Chicago Business. Chicago; 1988 September 19: 2.

Fantel, H. A plan to improve the picture on cable TV. In: New York Times. New York; 1987 September 13; sec. H: 44.

Farhi, P. Columbia acceptance of bid puts Sony into the big time. In: Washington Post. Washington, DC; 1989 September 28; sec. E: 1.

Farhi, P. Hollywood in the '80s: a boffo business rolls on. In: Washington Post. Washington, DC; 1989 March 26; sec. H: 1.

Farnsworth, C.H. The Bush team has competing ideas on competing with Japan. In: New York Times. New York; 1989 June 25; sec. E: 4.

Feldman, L. Digital TV - how soon? Computers & Electronics; 1983 September; 21: 85-89.

Feldman, L. High definition tele-vision. Radio-Electronics; 1989 February; 60: 33-38.

Feldman, L. I want my HDTV. Video Review; 1989 May; 10: 92.

Feldman, L. Improved definition tele-vision: while we wait for high-definition TV - there's improved-definition TV. Radio-Electronics; 1989 January; 60: 43-47.

Feldt, T.E. High-definition-TV scheme exploits inability of eye to detect all details of moving images. Electronics; 1984 May 17; 57: 48-49.

Fenton, B.C. HDTV update. Radio-Electronics; 1988 January; 59: 16-18.

Flanigan, J. Biggest failure pushed Sony to latest success. In: Los Angeles Times. Los Angeles; 1989 September 28; sec. 4: 1.

Flanigan, J. Catching up in TV technology won't be easy. In: Los
    Angeles Times. Los Angeles; 1988 January 18; sec. IV: 1.
Flanigan, J. Digital chip is real hope for video industry. In:
    Los Angeles Times. Los Angeles; 1989 May 14; sec. IV: 1.
Flanigan, J. Paying the price of indolence. Financial World; 1988
    June 28; 157: 103.
Flanigan, J. U.S. television must tune back in. In: Los Angeles
    Times. los Angeles; 1988 May 15; sec. IV: 1.
Flatow, I. Shootout at the HDTV corral. In: Los Angeles Times.
    Los Angeles; 1989 January 22; sec. IV: 3.
Fleischmann, M. The letterbox advantage. Video Magazine; 1989
    July; 13: 21-23.
Florio, J. Let's swap FSX for HDTV. In: Christian Science Monitor.
    Boston; 1989 June 12: 18.
Forbes, M.J. We now have to pay a price for having been first. Forbes;
    1988 September 5; 142: 17.
Fox, B. A sharper picture. World Press Review; 1986 July; 33: 54- 55.
Free, J. Coming: sharper TV. Popular Science; 1987 January; 230:
    72-76.
Gaffney, C. Information at your fingertips. In: Mercer Business.
    Trenton, NJ; 1989 August: 52.
Gallagher, R.T. High-definition TV faces hurdles. Electronics; 1983
    June 16; 56; ISSN 81.
Gallagher, R.T. Standard TV quality improves to near high- definition.
    Electronics; 1983 June 16; 56: 81-82.
Galluzzo, T. HDTV revisited: coming sooner than you think. Modern
    Photography; 1989 March; 53: 62-63.
Galluzzo, T. High-definition TV: it's possible, but where does it really
    stand right now? Modern Photography; 1982 January; 46:
    51-54.
Galluzzo, T. The quality edge: better video on the way! Modern Pho-
    tography; 1988 September; 52: 28-29.
Galluzzo, T. True high definition: are we nearly there? Modern Pho-
    tography; 1988 December; 52: 66-68.
Gardner, D.L. Where one hand washes the other: technical agree-
    ments between corporations build winning advantages in the
    world's most competitive markets. Design News; 1989 July 3;
    45: 72-76.
Garr, D. Fine-tuning television technology of the '90s. In: Newsday.

Melville, NY; 1988 September 11: 83.

Gellene, D. Small firms vying for key roles. In: Los Angeles Times. Los Angeles; 1989 March 9; sec. IV: 1.

Gellene, D. Zenith, AT&T join to develop high-definition television. In: Los Angeles Times. Los Angeles; 1989 March1; sec.IV:1.

Gerber, E. The lines' share. In: Houston Post. Houston, TX; 1988 July 3; sec. E: 2.

Gigot, P.A. Time to turn off this HDTV program. In: Wall Street Journal. New York; 1989 August 25; sec. A: 8.

Gildner, G. Forget HDTV, it's already outmoded. In: New York Times. New York; 1989 May 28; sec. B: 2.

Gilmartin,, P.A. Lawmakers to press for legislation to boost U.S. high-definition TV role. Aviation Week & Space Technology; 1989 March 27; 130: 24.

Gladwell, M. HDTV link to let pathologists study tissue from miles away. In: Washington Post. Washington, DC; 1989 July 3; sec. E: 5.

Glenn, W.E.; Glenn, K.G. Let's not lose another market to Japan. In: New York Times. New York; 1988 March 6; sec. E: 3.

Goldberg, R. A rosy, rectangular view of the future. In: Wall Street Journal. New York; 1989 June 5; sec. A: 17.

Goldman, J.S. HDTV funding gets a bleak reception. In: Business Journal-San Jose. San Jose, CA; 1989 August 7: 1.

Gosch, J. CRTs boast a multitude of design improvements. Electronics Week; 1984 November 5; 57: 30-31.

Gosch, J. Two-channel broadcast gives high-quality 3-D TV. Electronics; 1986 January 20; 59: 26-7.

Green, T. AEA will prepare for 1990 with Austin board meeting. In: Austin Business Journal. Austin, TX; 1989 September 18: 9.

Greene, B. Phone companies pave way for fight over cable TV rights. In: Wichita Business Journal. Wichita, KS; 1989 August 21: 15.

Gress, T. Phone companies gird for crucial battle over cable TV. In: Kansas City Business Journal. Kansas City, MO; 1988 November 28: 21.

Grier, P. Pentagon arms suffer from high-tech gap. In: Christian Science Monitor. Boston; 1989 June 8; sec. 1: 7.

Grunbaum, R. Mathews pictures magic with his TV whiz-bangs. In: Business Journal-Sacramento. Sacramento, CA; 1987

October 5: 8.

Hall, A.; Port, O. Why high-tech teams just aren't enough. Business Week; 1989 January 30: 63.

Harris, P. Commerce secretary opposes fed govt. high-D grant for Sony. Variety; 1989 March 15; 334: 43.

Harris, P. Electronics industry group tells Congress it needs $1.45-bil for HD development. Variety; 1989 May 17; 335:60.

Harris, P. FCC moves a step closer to delivery of high definition TV. Variety; 1988 September 7; 332: 63.

Harris, P. High-D bill offers incentives to stimulate U.S. research. Variety; 1989 March 8; 334: 50.

Hawkins, W.J. Picture perfect. Popular Science; 1988 June; 232: 64-67.

Hawley, G. Fiber optics comes full circle. Network World; 1987 January 19; 4(3): 28.

Hecht, J. American defence fears trigger television drama. New Scientist; 1989 April 8: 38-41.

Helliwell, J. HDTV shows promise, but don't hold your breath. PC Week; 1989 May 29; 6: 20.

Herndon, K. Cox, Tribune Broadcasting form HDTV group. In: Atlanta Constitution. Atlanta, GA; 1988 February 23; sec. D: 3.

Herndon, K. New HDTV gives look into future: revolutionary technology delivers sharper picture. In: Atlanta Journal- Constitution. Atlanta; 1987 December 5; sec. C: 1.

Hiatt, F. High-definition TV: Japan's next success? In: Washington Post. Washington, DC; 1988 September 11; sec. H: 1.

Hillkirk, J. Pentagon is hoping $60M will sharpen TVs. In: USA Today. Washington, DC; 1988 December 20; sec. B: 1.

Hirrel, M.J. TV's next generation. In: Christian Science Monitor. Boston; 1989 February 7: 19.

Hoban, P. Mick meets HDTV. New York; 1987 August 24; 20: 31.

Holden, D. Top five valley annual reports. In: Business Journal- San Jose. San Jose, CA; 1989 February 13: 19.

Iversen, W.R. Advances in personal video could fire up a new market. Electronics; 1988 October; 61: 121-2.

Iversen, W.R. Digital circuits begin to show up in TV receivers, but stereo sound is the hottest new feature. ElectronicsWeek; 1984 October 15; 57: 48.

Iversen, W.R. High-definition TV is still on hold. Electronics; 1982

December 29; 55: 51-52.

Jacobs, C. Stop your bellyaching and start selling products to the Japanese. In: Los Angeles Business Journal. Los Angeles; 1989 August 7: 35.

Jaffe, A. HDTV report soothes US fears of being shut out of market place. Television/Radio Age; 1988 December 12; 36: 14.

Jaques, B. HDTV technology excites at Geneva Telecom Confab. Variety; 1987 November 11; 329: 51.

Johnstone, B. Standards of vision. Far Eastern Economic Review (Hong Kong); 1989 March 9; 143(10): 77.

Johnston, R. High-definition TV poses challenge for US. In: Christian Science Monitor. Boston; 1989 January 5: 7.

Kalish, D. Creative concepts: picture this. Marketing and Media Decisions; 1989 March; 24(3): 32-3.

Kamienski, M. Fiber-optic technology has arrived but it costs too much. In: Intercorp. Hartford, CT; 1989 March 17: 1.

Kaplan, C.S. Zenith Corp. downloads computers. In: Newsday. Melville, NY; 1989 October 3: 41.

Kendall, P. Inventor/entrepreneur Denyse DuBrucq aims for the big leagues. In: Business Journal-Milwaukee. Milwaukee, WI; 1989 July 3: 10.

Kenny, G. Get the picture! A guide to high-technology TV. Stereo Review; 1989 April; 54: 69-73.

Kerr, J. Fed mega-funds for HDTV? Dream on, AEA. Electronic Business; 1989 June 26; 15(13): 172.

Kilborn, P.T. Antitrust shift will be proposed to aid TV effort: Mosbacher details plan; allowing companies to work jointly on high-definition sets is seen as crucial. In: New York Times. New York; 1989 May 4; sec. A: 1.

Kilborn, P.T. Support seen for TV technology aid. In: New York Times. New York; 1989 May 5; sec. B: 3.

Kilborn, P.T. U.S. funds sought for advanced TV. In: New York Times. New York; 1989 May 10; sec. B: 1.

Kindel, S. Pictures at an exhibition. Forbes; 1983 August 1; 132: 137-9.

Kindel, S. A sharper image. Financial World; 1988 August 23; 157: 35-36.

Kindel, S. A sharper image. Financial World; 1988 August 23; 157: 35-37.

Kipps, C. In search for visual perfection, budget referees HD vs. film fight. Variety; 1988 October 5; 332: 95.

Knight, B. Made-for-TV high-D film is brought in on budget. Variety; 1989 March 29; 334: 35.

Kriz, M.E. Looking sharp: high-definition television, a new technology that Japan might soon start selling to American consumers, poses high-stakes challenges across the TV spectrum. National Journal; 1988 July 16; 20: 1860-3.

Kuttner, B. America's telecommunications blinders. In: Boston Globe. Boston; 1988 October 13: 15.

Kuttner, R. Why we don't make TVs anymore. In: Washington Post. Washington, DC; 1988 October 14; sec. A: 25.

Kuzela, L. Liquid-crystal projection will require no film. Industry Week; 1989 February 6; 238(46-47).

Lachenbruch, D. Buy American? A Silicon Valley-led consortium wants to get U.S. manufacturers back in the game with HDTV. Channels The Business of Communications; 1988 December; 8: 124.

Lachenbruch, D. Cutting through the high-definition hype. Video Magazine; 1989 March; 12: 118.

Lachenbruch, D. HDTV: picture perfect. Channels of Communications; 1985 November/December; 5: 77.

Lachenbruch, D. Picture this. TV Guide; 1989 April 15; 37: 19.

Lachenbruch, D. The push is on to deliver picture-perfect TV: competitors are racing to come up with a high-definition system that'll work in America. TV Guide; 1988 February 6; 36: 36-37.

Lachenbruch, D. Sharper and super: a new VCR's got the picture. Channels The Business of Communications; 1987 December; 7: 124-5.

Lachenbruch, D. You'll be seeing things more clearly. TV Guide; 1983 June 4; 31: 45-47.

Lachica, E. Small U.S. firms challenge Japanese grip on HDTV. In: Wall Street Journal. New York; 1989 October 23; sec. B: 1.

Lander, D. Third-generation television: hi-fi for your eyes and ears. Popular Mechanics; 1985 February; 162: 82-85.

Lazare, L. WTTW tunes in new funding sources. In: Crains Chicago Business. Chicago; 1988 December 5: 3.

LeDuc, D. Local firm producing high-definition TV show. In:

Nashville Business Journal. Nashville, TN; 1989 April 3: 16.

Lee, M. Southwestern Bell president seeks technology without rate increase. In: Tulsa World. Tulsa, OK; 1989 August 29; sec. B: 5.

Leib, J. Colorado wins out as headquarters site for cable research lab. In: Denver Post. Denver, CO; 1989 January 13; sec. C: 1.

Leopold, G.; Gallagher, R. Disputes may stall HDTV standard for two years. Electronics; 1986 May 12; 59: 19-20.

Levine, M. An inventor toils in Westbury and the Pentagon hopes for a breakthrough. In: Newsday. Melville, NY; 1989 July 3; sec. 3: 2.

Levine, M. Make HDTV a public-private venture. In: New York Times. New York; 1988 November 28; sec. A: 24.

Levis, A. Straight talk on HDTV. Video Magazine; 1989 August; 13: 6.

Levy, S. Next picture show. Rolling Stone; 1989 June 15: 91-97.

Lewis, P.H. Advances in television; picture is brighter, its future is murky. In: New York Times. New York; 1987 April 8; sec. D: 8.

Lewyn, M. FCC sets rules for high definition TV. In: USA Today. Washington, DC; 1988 September 2; sec. B: 1.

Lewyn, M. High-definition TV: a chance to regain lead. In: USA Today. Washington, DC; 1988 May 2; sec. E: 1.

Lewyn, M. Techtalk. In: USA Today. Washington, DC; 1988 September 8; sec. B: 6.

Lindsey, J. Manufacturing sector takes lead role in county economy. In: Boulder County business Report. Boulder, CO; 1989 February: 17.

Lipman, J. Firms use high-definition ads but viewers can't see effects. In: Wall Street Journal. New York; 1989 March 23; sec. B: 7.

Lippman, J. High definition TV picture still fuzzy: Japan is gaining on technology. Variety; 1988 April 13; 330: 45-46.

Litvan, L.M. High-definition TV firms look to the Pentagon.In: Washington Post. Washington, DC; 1989 September 15; sec. C: 1.

Lois, T. The HDTV picture starts coming into focus: the FCC sets initial technical guidelines for high-definition TV. Business Week; 1988 September 19: 38.

Magnusson, P. Promoting high-definition TV: the perils for Uncle

Sam. Business Week; 1989 May 27: 30.

Marbach, W.D. Geometry could give HDTV signals the right shape. Business Week; 1989 April 3: 110.

Marbach, W.D. A report from the Hill gives HDTV a weak reception. Business Week; 1989 August 14: 89.

Marbach, W.D. et al. Super television: the high promise - and high risks - of high-definition TV. Business Week; 1989 January 30: 56-61.

Marbach, W.D. Where the jobs will be in high-definition TV - overseas. Business Week; 1988 May 2: 99.

Margasak, L. FCC approval of 'High Definition' TV praised. In: Los Angeles Times. Los Angeles; 1988 September 3; sec. V: 10.

Markey, E.J. Pendulum swinging back from blind deregulation toward protecting the public. Variety; 1988 January 20; 329: 181.

Markham, J.M. NBC proposes standard for a better TV picture. In: New York Times. New York; 1988 October 17; sec. D: 2.

Markoff, J. NBC proposes TV standard for high-quality pictures. In: New York Times. New York; 1988 October 17; sec. D: 2.

Marshall, P.G. A high-tech, high-stakes HDTV gamble. Editorial Research Reports; 1989 February 17: 90-103.

Marson, C. Sanders has parlayed experience and capital into leading video house. In: Indianapolis Business Journal. Indianapolis, IN; 1989 January 9: 8.

Mason, C. Excited federal lawmakers herald HDTV era. Telephony; 1988 September 12: 8-9.

Mason, C. FCC zooms in in HDTV. Telephony; 1988 September 9: 3.

Matsumoto, N. High-definition TV has fuzzy future. In: San Francisco Chronicle. San Francisco; 1988 August 7; sec. D: 1.

McWalter, K. HDTV: turning point for television: high- definition systems promise sharper-than-ever TV images, but the broadcast industry must bring standards into focus. Design News; 1988 May 23; 44: 22-23.

Michael, P. Chasing rainbows. Cinema Canada; 1986 September: 5.

Middleton, A. Cabot regroups in San Diego, hunts new market. In: San Diego Business Journal. San Diego, CA; 1989 June 26: 3.

Mitchell, P.W. High-tech video. The Atlantic; 1985 December; 256: 102-3.

Monaco, J. Into the '90s. American Film; 1989 January/February;

14: 24-27.

Moran, T. Network TV is changing, but won't bow to cable. In: Crains Detroit Business. Detroit, MI; 1989 July 31; sec. 2: E-4.

Mowrer, W. Hi-fi for your eyes! High Fidelity; 1984 May; 34: 41- 43.

Mullen, N. High-definition TV: tiny organisms seem close. In: Christian Science Monitor. Boston; 1988 May 31: 16.

Myerson, A.R. The odd couple. In: Georgia Trend. Atlanta, GA; 1988 September: 46.

Nadan, J.S. A glimpse into future television: a technology evolving in parallel with personal computers. Byte; 1985 January; 10: 135-45.

Naegele, T. U.S. broadcasters seek an entree to HDTV. Electronics; 1987 January 8; 60: 33-34.

Naegele, T. Will the U.S. be a follower in HDTV technology? Electronics; 1988 September; 61: 33-34.

Norman, C. A focus on advanced television? Science; 1989 April 14; 244: 137.

Norman, C. HDTV: the technology du jour. Science; 1989 May 19; 244: 781-4.

Norris, E. Zenith passes milestone in high-resolution TV race. In: Crains Chicago Business. Chicago, Il; 1988 November 7: 41.

Pae, P. Kodak enters the HDTV market with converter for movie film. In: Wall Street Journal. New York; 1989 October 23; sec. B: 4.

Palmer, J. Brightening picture: for Zenith's fortunes, the nadir is past. Barron's; 1989 March 20; 69(12): 13, 40.

Pappas, C.L. Light reading: future shock in the fiber optics industry. In: Business Worcester. Worcester, MA; 1988 January 11: 14.

Passell, P. High definition: TV battleground. In: New York Times. New York; 1988 November 9; sec. D: 2.

Passell, P. Sharp TV images a complex topic. In: New York Times. New York; 1988 November 16; sec. D: 2.

Pine, A. U.S. seeks ways to win race for highly defined TV. In: Los Angeles Times. Los Angeles; 1989 May 4; sec. I: 1.

Point, O.; Armstrong, L.; Gross, N. High-definition TV is rallying a digital revolution. Business Week; 1989 January 30: 64-6.

Poletti, T. High definition TV: Japan pushes ahead while U.S. industry debates strategy. In: Business Journal-San Jose. San Jose, CA; 1989 January 30: 10.

Pollack, A. Electronics concerns study TV development venture. In: New York Times. New York; 1989 January 13; sec. D: 2.

Pollack, A. U.S.-Europe technology urged. In: New York Times. New York; 1989 July 24; sec. D: 1.

Pool, R. Setting a new standard. Science; 1988 October 7; 242: 29-31.

Pool, R. Super tube: here comes high-tech TV. In: Washington Post. Washington, DC; 1988 December 18; sec. A: 1.

Port, O. HDTV: Washington still isn't receiving the signal. Business Week; 1988 November 21: 114.

Port, O. Resolved: the future of TV-screen resolution will be put off. Business Week; 1986 June 23: 133.

Port, O.; Armstrong, L.; Gross, N. High-definition TV is rallying a digital revolution: the technology is causing many industries to blend - and Japan is poised to take the lead. Business Week; 1989 January 30: 64-66.

Porter, W. CBO report on high-definition TV gets panned on Hill. In: Washington Times. Washington, DC; 1989 August 2; sec. C: 5.

Postrel, V.I. Sharper images. Reason; 1989 April; 20: 8-9.

Powell, A. Broadcaster tops rival, faces new competitors. In: Intercorp. Hartford, CT; 1988 July 10: 24.

Ranada, D. Aspects of HDTV. High Fidelity; 1989 April; 39: 47-50.

Ranada, D. Seeing the future. High Fidelity; 1987 August; 37: 22.

Redburn, T. Commerce Secretary cool to HDTV funding; research by electronics industry shouldn't depend on federal money, Mosbacher says. In: Los Angeles Times. Los Angeles; 1989 May 10; sec. IV: 7.

Redburn, T. Difference between 'us' and 'them' blurs in a global economy. In: Los Angeles Times. Los Angeles; 1989 August 8; sec. IV: 1.

Redburn,T. Huge Pentagon grants sought to develop HDTV. In: Los Angeles times. Los Angeles; 1989 May 9; sec. IV: 1.

Redburn, T. Sony taps Silicon Valley for share of HDTV research. In: Los Angeles Times. Los Angeles; 1989 April 28; sec. IV: 1.

Redburn, T. U.S. names 5 firms to receive grants for high- definition TV. in: Los Angeles Times. Los Angeles; 1989 June 14;

sec.IV: 1.

Reneteau, P.J.; McCloskey, P.F. With HDTV, America can reclaim its lead in electronics. In: New York Times. New York; 1988 November 18: 26.

Revzin, P. Europeans gamble on movie-quality TV: Japanese effort isn't certain; lesson for U.S.? In: Wall Street Journal. New York; 1988 September 30: 22.

Richards, E. Computer plan a boost for companies. In: Washington Post. Washington, DC; 1989 September 10; sec. H: 3.

Richards, E. Consortia: the new business cure-all? In: Washington Post. Washington, DC; 1989 May 26; sec. D: 10.

Richards, E. Doubting the focus on HDTV: critics say U.S. industry may be making costly error. In: Washington Post. Washington, D.C.; 1989 May 21; sec. H: 1.

Richards, E. Focus sharpens on high-definition TV. In: Washington Post. Washington, DC; 1989 February 2; sec. E: 1.

Richards, E. HDTV provides spark of hope for failing Silicon Valley firm - and U.S. electronics industry. In: Washington Post. Washington, DC; 1989 June 15; sec. E: 1.

Richards, E. HDTV's prospects oversold, Congressional study says. In: Washington Post. Washington, DC; 1989 July 28; sec. E: 1.

Richards, E. Homeward look tests high-tech companies. In: Washington Post. Washington, DC; 1989 June 30; sec. D: 1.

Richards, E. Pentagon aims to revive U.S. TV industry. In: Washington Post. Washington, DC; 1988 December 19; sec. A: 1.

Richards, E. Soviets offer to 'referee' HDTV race. In: Washington Post. Washington, DC; 1989 February 23; sec. E: 1.

Richards, E. Study sees Sematech as successful so far. In: Washington Post. Washington, DC; 1989 May 9; sec. B: 1.

Richards, E. 'Superhighway' for data could speed U.S. to top in science, technology. In: Los Angeles Times. Los Angeles; 1989 September 7; sec. 4: 1.

Richards, E. Zenith computer unit sold to French company. In: Washington Post. Washington, DC; 1989 October 3; sec. C: 1.

Richter, P. Broadcasters give new TV system static. In: Los Angeles Times. Los Angeles; 1987 October 11; sec. IV: 1.

Riley, K. Electronics firms seek federal aid. In: Washington

Times. Washington, DC; 1989 May 10; sec. C: 1.

Rivlin, R. High-def perplex. Channels The Business of Communications; 1988 January; 8: 78.

Robertson, J. Falling flat on HDTV displays. Electronic News; 1989 June 19; 35(1763): 13.

Rodrian, S. Layoffs, setbacks don't quash optimism of Cal Micro's chief. In: Business Journal-Phoenix & the Valley of the Sun. Phoenix, AZ; 1989 March 20: 8.

Rogers, M. The television of the future: the United States, Japan and Europe are racing to build bigger, better and much more expensive sets. Newsweek; 1988 April 4; 111: 62-63.

Rooney, P. DARPA spends less than $30M on five HDTV hopefuls. EDN; 1989 July 13; 34(13A): 5.

Rosenblatt, R.A. Industry asks U.S. for help in developing new TV sets. In: Los Angeles Times. Los Angeles; 1988 November 6; sec. IV: 1.

Rosenblatt, R.A. Networks urge slow shift to sharper TV picture. In: Los Angeles Times. Los Angeles; 1988 June 24; sec. 4: 1.

Ross, C. Cable giants back high-definition TV. In: San Francisco Chronicle. San Francisco; 1988 May 2; sec. F: 1.

Roth, C. New videotape system promises to improve TV picture quality. In: New York Times. New York; 1988 April 13; sec. D: 6.

Roth, M. HDTV: national but still small. Variety; 1988 April 20; 330: 1.

Rowan, H. High-density TV; another Japanese marketing success in the making. In: Washington Post. Washington, DC; 1987 September 24; sec. A: 25.

Rowan, H. Should U.S. be focusing on HDTV? In: Washington Post. Washington, DC; 1989 March 26; sec. H: 1.

Rowe, J. From Howdy Doody to HDTV. In: Christian Science Monitor. Boston; 1989 June 14: 12.

Saffo, P. Multimedia: seeing is deceiving. Personal Computing; 1989 August; 13(8): 181-2.

Salibian, E.C. R&D spending likely to remain high. In: Rochester Business Journal. Rochester, NY; 1989 January 23: 17.

Samuelson, R.J. HDTV high-tech pork barrel. In: Washington Post. Washington, DC; 1989 May 17; sec. A: 23.

Sanger, D.E. Japan begins first regular broadcasts of television of the future. In: New York Times. New York; 1989 June 4; sec. 1: 13.

Sanger, D.F. Japanese test illustrates big lead in TV of future. In: New York Times. New York; 1989 March 21; sec. A: 1.

Schenker, J. How Thomson aims to build a world-class TV business: but the Japanese could upset those plans with HDTV. Electronics; 1987 October 29; 60: 57-58.

Schoenberger, K. Japan touting HDTV despite international static. In: Los Angeles Times. Los Angeles; 1989 September 25; sec. 4: 1.

Schrage, M. Is it a TV or a PC ? In: Washington Post. Washington, DC; 1988 December 18; sec. C: 3.

Schreiber, P. Cablevision takes big bites. In: Newsday. Melville, NY; 1988 August 19; sec. 3: 1.

Schreiber, P. Shots ring out in phone war. In: New York Newsday. New York; 1989 February 14; sec. 1: 37.

Schwartz, J. Putting America in the picture: trailing Japan badly, U.S. firms ask Washington for help in the race to build high-definition TV. Newsweek; 1989 May 29; 113: 42-43.

Schwartz, R.S. HDTV: keeping the rabbit ears alive. High Fidelity; 1989 April; 39: 51-53.

Sedore, D. FAU grabs lead role in research work on high- definition TV. In: South Florida Business Journal. Miami, FL; 1989 July 24: 1.

Segers, F. High definition TV bursting out of laboratories into theaters. Variety; 1987 October 14; 328: 158.

Seghers, F. Television makers are dreaming of a wide crispness. Business Week; 1987 December 21: 108-9.

Sharbutt, J. CBS movie tests new TV technology. In: Los Angeles Times. Los Angeles; 1988 March 28; sec. VI: 1.

Shear, J. Computer TV may help U.S. zoom past world competitors. In: Washington Times. Washington, DC; 1989 March 30; sec. C: 1.

Shear, J. To program TVs for a new market. Insight; 1989 March 20; 5: 40-41.

Shear, J. Zenith warms up for new set in the American TV turf war. Insight; 1989 April 24; 5: 40.

Shulman, S. Not a pretty picture. The Progressive; 1989 September; 53(9): 24-5.

Sims, C. F.C.C. sets technical guidelines for high-definition TV in 1990's. In: New York Times. New York; 1988 September 2; sec. A: 1.

Sims, C. Five to get U.S. grants for advanced TV. In: New York Times. New York; 1989 June 14; sec. D: 7.

Sims, C. HDTV: will U.S. be in the picture? Consortium sought to aid development. In: New York Times. New York; 1988 September 21: 27.

Sims, C. Hearings on high-definition TV. In: New York Times. New York; 1989 March 9; sec. D: 1.

Sims, C. Sony seeks TV grant from U.S. In: New York Times. New York; 1989 March 3; sec. D: 1.

Sims, C. Striving to keep its cutting edge: Sarnoff Research Center tackles HDTV - and life after RCA. In: New York Times. New York; 1989 April 30; sec. B: 4.

Sims, C. A.T.&T. and Zenith in TV deal. In: New York Times. New York; 1989 March 1; sec. D: 1.

Sims, C. U.S. researchers show gains in the television of the future. In: New York Times. New York; 1989 April 16; sec. A: 1.

Sims, C. U.S. warned to be strong in sharp TV. In: New York Times. New York; 1988 November 23; sec. B: 1.

Sisler, S. What's happening in books, TV, and viruses? Online; 1988 July; 12: 103-6.

Siwolop, S. High-definition pix may first hit the stix. Business Week; 1988 June 6: 99.

Slutsker, G. Goodbye cable TV, hello fiber optics. Forbes; 1988 September 19; 142: 174-79.

Smith, E.T. Keeping your TV off the endangered species list. Business Week; 1988 March 21: 151.

Smith, E.T. Sony expands its HDTV work - and asks Uncle Sam to chip in. Business Week; 1989 May 22: 123.

Smith, L. Can consortiums defeat Japan? Fortune; 1989 June 5; 119: 245-9.

Smith, L.M. EIA founds HDTV resource center: unveils HDTV study. Back Stage; 1988 December; 29: 1-4.

Solomon, D. HDTV: television's next generation. Marketing and Media Decisions; 1989 March; 24: 109-111.

Somerfield, H. Television of tomorrow: high-definition magic.

In: San Francisco Chronicle. San Francisco; 1988 February 10; sec. BRI: 10.

Stoddard, A.C.; Dibner, M.D. Europe's HDTV: tuning out Japan. Technology Review; 1989 April; 92: 39.

Sukow, R.; Jessell, H. High definition TV. Broadcasting; 1989 May 15; 116: 38-9.

Tannenbaum, J.A. Picture perfect? Next generation of TVs promises breakthrough in image quality. In: Wall Street Journal. New York; 1987 August 24: 25.

Tannenbaum, J.A. Small pioneers in HDTV bet on a head start. In: Wall Street Journal. New York; 1989 August 1; sec. B: 1.

Thomas, E., Jr. High-definition TV is coming; question is when. In: Atlanta Business Chronicle. Atlanta, GA; 1988 April 25; sec. B: 3.

Thorpe, L. HDTV: it's the size that counts. Video Review; 1989 August; 10: 92.

Tyson, D.A. U.S. competitiveness and new TVs. In: Los Angeles Times. Los Angeles; 1989 February 5; sec. IV: 2.

Vartabedian, R. Hughes wins $1-Billion deal for 11 satellites. In: Los Angeles Times. Los Angeles; 1988 July 30; sec. 4: 1.

Waldman, P. AT&T and Zenith team up to produce, sell high-definition TVs in early 1990s. In: Wall Street Journal. New York; 1989 March 1; sec. B: 4.

Waldman, P. High-definition TV sparks controversy at electronics show. In: Wall Street Journal. New York; 1989 January 10; sec. B: 5.

Waldman, P. Sixteen U.S. companies to form groups to develop advanced-TV technology. In: Wall Street Journal. New York; 1989 January 13; sec. A: 2.

Walker, S.L. The contractor that's coming in from the cold. Business Week; 1988 November 14: 150-1.

Waller, L. Fiber's new battleground: closing the local loop. Electronics; 1989 February; 62: 94-96.

Warren, R. Disc, DAT and other things. Playboy; 1988 June; 35: 106-10.

Weber, D.M. Digital circuits point towards better TV sets. ElectronicsWeek; 1984 August 13; 57: 49-53.

Weber, D.M. The drive to sharpen NTSC TV picture begins. Electronics; 1985 December 23; 58: 59-60.

Weber, D.M. High-definition TV waits in wings for demand to build: manufacturers have prototypes, but seek go-ahead form broadcasters before trying to interest the public. ElectronicsWeek; 1985 June 17; 58: 35-36.

Weber, J.; Richter, P. Joint venture in memory chips proposed by 7 U.S. companies. In: Los Angeles Times. Los Angeles; 1989 June 22; sec. 4: 1.

Weiss, J.M.; Horino, M. Advanced TV's lessons for US R&D. In: Christian Science Monitor. Boston; 1988 November 8: 14.

Welch, R. In Kansas: telecommunications on a roll. In: Kansas Business News. Lindsborg, KS; 1989 February: 18.

Wessel, D.; Lachica, E. Mosbacher's initiative on HDTV is getting scuttled, sources say. In: Wall Street Journal. New York; 1989 August 2; sec. B: 2.

Wetmore, T. The HDTV debate. Channels The Business of Communications; 1988 November; 8: 86.

Wetmore, T. Some tough choices: can producers afford emerging technologies? Can they afford to ignore them? Channels The Business of Communications; 1989 March; 9: 63.

Whyte, B. Too hot to handle. Audio; 1989 September; 73: 24-25.

Wiegner, K.K. Last chance? Forbes; 1988 May 30; 141: 58-60.

Wielage, M. HDTV, new gear, steal show for broadcasters. Video Review; 1989 July; 10: 15.

Wielage, M. The newest and the best in home entertainment: hot new products - digital audiotape players and Super-VHS VCRs - promise the best in sight and sound enjoyment. Consumers Digest; 1988 January/February; 27: 26-29.

Williams, S. Best bet? In: Corporate Report Minnesota. Minneapolis, MN; 1989 July: 33.

Wilson,C. Bellcore devises digital coding for HDTV. Telephony; 1988 September 9: 8-9.

Wilson, J. Better way to train engineers. In: Focus. Philadelphia, PA; 1988 December 7: 3.

Winkler, E. Call HDTV boon in gear, materials. Electronic News; 1989 July 10; 35(1766): 39-40.

Winner, L. Who needs HDTV? Technology Review; 1989 May/June; 92: 20.

Winston, B. The built-in bias against HDTV. In: New York Times. New York; 1989 April 16; sec. B: 3.

Winston, B. The US and HDTV. In: Christian Science Monitor.
    Boston; 1988 December 2: 15.
Winter, C. Turning on to high-tech TV no simple matter. In: Chicago
    Tribune. Chicago; 1988 July 4; sec. 3: 1.
Wolf, J. Europeans fear obstacles by U.S. on advanced TV. In: Wall
    Street Journal. New York; 1989 May 31; sec. A: 14.
Wood, C. Cable labs to test HDTV at local sites. In: Boulder County
    Business Report. Boulder, CO; 1989 June: 6.
Wood, C. Cable TV labs picks Boulder for interim site. In: Boulder
    County Business Report. Boulder, CO; 1989 March: 25.
Young, B. The new look: high-definition television promises to
    change the face of home entertainment. Rolling Stone; 1988
    October 20: 95-96.
Young, J.; Cohen, C.L.; Iversen, W.R. At last, the tv picture gets
    sharper. Electronics; 1987 October 15; 60: 113-4.